HEALTHINFO 2018

The Third International Conference on Informatics and Assistive Technologies for Health-Care, Medical Support and Wellbeing

October 14 - 18, 2018

Nice, France

HEALTHINFO 2018 Editors

Hassan Khachfe, Lebanese International University, Lebanon

Birgit Gersbeck-Schierholz, Leibniz Universität Hannover, Germany

Printed from e-media with permission by:

Curran Associates, Inc.
57 Morehouse Lane
Red Hook, NY 12571

Some format issues inherent in the e-media version may also appear in this print version.

Printed by Curran Associates, Inc. (2018)

International Academy, Research, and Industry Association (IARIA)
412 Derby Way
Wilmington, DE 19810

Phone: (408) 893-6407
Fax: (408) 527-6351

petre@iaria.org

Additional copies of this publication are available from:

Curran Associates, Inc.
57 Morehouse Lane
Red Hook, NY 12571 USA
Phone: 845-758-0400
Fax: 845-758-2633
Email: curran@proceedings.com
Web: www.proceedings.com

Table of Contents

Souza Coelho Soares

HEALTHINFO 2018

Forward

The Third International Conference on Informatics and Assistive Technologies for Health-Care, Medical Support and Wellbeing (HEALTHINFO 2018), held on October 14 - 18, 2018- Nice, France, tackles with particular aspects belonging to health informatics systems, health information, health informatics data, health informatics technologies, clinical practice and training, and wellbeing informatics in terms of existing and needed solutions.

The progress in society and technology regarding the application of systems approaches information and data processing principles, modeling and information technology, computation and communications solutions led to a substantial improvement of problems in assistive healthcare, public health, and the everyday wellbeing. While achievements are tangible, open issues related to global acceptance, costs models, personalized services, record privacy, and real-time medical actions for citizens' wellbeing are still under scrutiny.

We take here the opportunity to warmly thank all the members of the HEALTHINFO 2018 technical program committee as well as the numerous reviewers. The creation of such a broad and high quality conference program would not have been possible without their involvement. We also kindly thank all the authors that dedicated much of their time and efforts to contribute to the HEALTHINFO 2018. We truly believe that thanks to all these efforts, the final conference program consists of top quality contributions.

This event could also not have been a reality without the support of many individuals, organizations and sponsors. We also gratefully thank the members of the HEALTHINFO 2018 organizing committee for their help in handling the logistics and for their work that is making this professional meeting a success.

We hope the HEALTHINFO 2018 was a successful international forum for the exchange of ideas and results between academia and industry and to promote further progress in health informatics research. We also hope Nice provided a pleasant environment during the conference and everyone saved some time for exploring this beautiful city.

HEALTHINFO 2018 Steering Committee

Enrico Vicario, University of Florence, Italy
Wendy MacCaull, St Francis Xavier University, Canada
Chun-Hsi Huang, University of Connecticut, USA
Toshiyo Tamura, Waseda University | Future robotics organization, Japan
Saif Khairat, University of North Carolina - Chapel Hill, USA

HEALTHINFO 2018 Researsch/Academia Advisory Committee

Dieter Hayn, AIT Austrian Institute of Technology, Austria
Eleanor Horton, School of Nursing, Midwifery and Paramedicine | University of the Sunshine Coast, Australia
Doris Lindörfer, Ludwig-Maximilians-Universität München | IBE - Institut für medizinische

Informationsverarbeitung, Biometrie und Epidemiologie, Germany
David Greenhalgh, University of Strathclyde, UK

HEALTHINFO 2018

Committee

HEALTHINFO Steering Committee
Enrico Vicario, University of Florence, Italy
Wendy MacCaull, St Francis Xavier University, Canada
Chun-Hsi Huang, University of Connecticut, USA
Toshiyo Tamura, Waseda University | Future robotics organization, Japan
Saif Khairat, University of North Carolina - Chapel Hill, USA

HEALTHINFO Researsch/Academia Advisory Committee
Dieter Hayn, AIT Austrian Institute of Technology, Austria
Eleanor Horton, School of Nursing, Midwifery and Paramedicine | University of the Sunshine Coast, Australia
Doris Lindörfer, Ludwig-Maximilians-Universität München | IBE - Institut für medizinische Informationsverarbeitung, Biometrie und Epidemiologie, Germany
David Greenhalgh, University of Strathclyde, UK

HEALTHINFO 2018 Technical Program Committee

Emmanuel Agu, Worcester Polytechnic Institute, USA
Jalel Akaichi, University of Tunis, Tunisia
Jesús B. Alonso Hernández, Institute for Technological Development and Innovation in Communications (IDeTIC) | University of Las Palmas de Gran Canaria (ULPGC), Spain
J. Antão B. Moura, Federal University of Campina Grande (UFCG), Brazil
Marjan Askari, Universiteit Utrecht, Netherlands
Panagiotis D. Bamidis, Aristotle University of Thessaloniki, Greece
Hugo Barbosa, Lusofona University of Porto / Faculty of Engineering of the University of Porto, Portugal
Rodolfo Barriviera, Instituto Federal do Paraná, Brazil
Kawtar Benghazi, University of Granada, Spain
Manuel Campos Martínez, University of Murcia, Spain
Chinmay Chakraborty, Birla Institute of Technology, Jharkhand, India
Po-Hsun Cheng, National Kaohsiung Normal University, Taiwan
Kevin Bretonnel Cohen, University of Colorado - School of Medicine, USA
Vassilis Cutsuridis, University of Lincoln, UK
Michael Czaplik, University Hospital RWTH Aachen, Germany
Arianna Dagliati, The University of Manchester, UK
Giuseppe De Pietro, Institute for High Performance Computing and Networking (ICAR) - Italian National Research Council (CNR), Italy / Temple University's College of Science and Technology, Philadelphia, USA
Steven A. Demurjian, The University of Connecticut, USA
Spiros Denaxas, University College London, UK
Anatoli Djanatliev, University of Erlangen-Nuremberg, Germany
Tunca Dogan, METU, Ankara, Turkey / EMBL-EBI, Cambridge, UK
António Dourado, University of Coimbra, Portugal

Peter Elkin, University at Buffalo, USA
Mechili Enkeleint-Aggelos, University of Vlora, Albania
Elif Eryilmaz, Technische Universitaet Berlin, Germany
Lincoln Faria da Silva, Institute of Computing - Fluminense Federal University, Niterói, Brazil
Paulo Freire Sobrinho, Federal University of Grande Dourados, Brazil
Sebastian Fudickar, University Oldenburg, Germany
Juliano Gaspar, Federal University of Minas Gerais, Brazil
Chris Gniady, The University of Arizona, USA
Adrian Gomez, Hospital Italiano de Buenos Aires, Argentina
Adela Grando, Arizona State University, USA
Raffaele Gravina, University of Calabria, Italy
David Greenhalgh, University of Strathclyde, UK
Irene Y. H. Gu, Chalmers University of Technology, Sweden
Udo Hahn, Jena University Language & Information Engineering (JULIE), Germany
Dieter Hayn, AIT Austrian Institute of Technology, Austria
José Luis Hernández Cáceres, Institute for Neurology and Neurosurgery Havana, Cuba
Harry Hochheiser, University of Pittsburgh, USA
Eleanor Horton, School of Nursing, Midwifery and Paramedicine | University of the Sunshine Coast, Australia
Chun-Hsi Huang, University of Connecticut, USA
Ming Huang, Mayo Clinic, USA
Michal Huptych, Czech Institute of Informatics, Robotics, and Cybernetics (CIIRC) | Czech Technical University (CTU) in Prague, Czech Republic
Fábio Iaione, Universidade Federal de Mato Grosso do Sul, Brazil
Zbigniew Kalbarczyk, University of Illinois at Urbana-Champaign, USA
Saif Khairat, University of North Carolina - Chapel Hill, USA
Eizen Kimura, Medical School of Ehime University, Japan
Evdokimos I. Konstantinidis, Nively Sas, Nice, France / Aristotle University of Thessaloniki, Greece
Constantinos Koutsojannis, TEI of Western Greece, Greece /
Francine Krief, Bordeaux INP, France
Tomohiro Kuroda, Kyoto University / Kyoto University Hospital, Japan
Valderi Leithardt, University of Vale do Itajai - Univali, Brazil
Doris Lindörfer, Ludwig-Maximilians-Universität München | IBE - Institut für medizinische Informationsverarbeitung, Biometrie und Epidemiologie, Germany
Guillermo Hugo Lopez Campos, Queen's University Belfast, UK
Wendy MacCaull, St Francis Xavier University, Canada
Anthony Maeder, Flinders University, Australia
Flora Malamateniou, University of Piraeus, Greece
Fernando J Martin-Sanchez, Weill Cornell Medicine | Cornell University, USA
Miguel A. Mayer, Universitat Pompeu Fabra, Spain
James P. McGlothlin, Fusion Consulting Inc, USA
Paolo Melillo, University of Campania Luigi Vanvitelli, Italy
Majid Mirmehdi, University of Bristol, UK
Hans Moen, University of Turku, Finland
Fadi Muheidat, The University of the Pacific, Stockton, California, USA
Tadashi Nakano, Osaka University, Japan
António Nogueira, University of Aveiro / Instituto de Telecomunicações, Portugal
Nelson Pacheco Rocha, University of Aveiro, Portugal

Nadine Vigouroux, IRIT | Université Paul Sabatier, France
Massimo Villari, Universita' di Messina, Italy
Urjaswala Vora, Centre for Development of Advanced Computing (C-DAC), Mumbai, India
Konstantinos Votis, Information Technologies Institute | Centre for Research and Technology Hellas, Greece
Klemens Waldhör, FOM Hochschulzentrum Nürnberg, Germany
Shouyi Wang, University of Texas at Arlington, USA
Wenwu Wang, University of Surrey, UK
Yanshan Wang, Mayo Clinic, USA
Sule Yildirim-Yayilgan, Norwegian University of Science & Technology, Norway
Huiru (Jane) Zheng, Ulster University at Jordanstown, UK

Holistic Capability Model for Sustainable Evolution of Health Care Providers

Michael Georg Grasser, Erich Sorantin
Research Unit of Digital Information and Image Processing
Medical University Graz
Graz, Austria
e-mail: michael.grasser@medunigraz.at,
erich.sorantin@medunigraz.at

Viktoria Schlaipfer, Robert Priedl
IT-Operations
Hospitaller Order of Saint John of God
Graz, Austria
e-mail: viktoria.schlaipfer@bbito.at, robert.priedl@bbito.at

Abstract—**Several factors have significantly influenced the digital evolution of health care providers worldwide. Existing consulting models and decision-making strategies are stretched to their limits. To derive sustainable decisions in a holistic manner an innovative digital capability model had to be developed and enforced. On that basis human beings, e.g., medical or nursing staff, are centered in the digital transformation. We have developed a holistic transformation model to successfully support the digital transformation processes of health care providers.**

Keywords-Holistic Capability Model; Human Based Digital Intelligence; Digital Transformation; Sustainable Decision-Making for Health Care Providers.

I. INTRODUCTION

In the whole wide world, hospitals have to compete with each other regarding quality of treatment or exclusiveness of their facilities that patients decide for them. Efficient operation and furthermore sustainable evolution of health care providers like hospitals, retirement homes or care centers, will be influenced and challenged by many factors from their ecosystems. Digitalization is currently one of the biggest challenges, beside the cost cut in public funding or the implementation of the general data protection regulation. For health care providers, it is necessary to develop and optimize their organizations in a digital manner to strategically adapt existing business models.

However, to make sustainable decisions in this fast changing and highly technological environment is difficult. This results in a great demand to use innovative and multifactorial decision-making methods. Our research focused on the development of a new holistic model to engage digital transformation processes. The results can be used as templates for several health care providers worldwide.

In section II we discuss related work and dissociate several models from our approach. A detail explanation of our digital transformation model as well as a transition to a technology from aeronautics is done in section III. Further fields of research and a conclusion can be found in section VI.

II. RELATED WORK

Considerations in digitalization center primarily on technology. So is there any chance to make sustainable decisions just by looking at technology? Or, the opposite, can one do digitalization completely without technology? For health care providers, it is a fact that they need more efficient technology to support the clinical pathways. Is technology the magic bullet for sustainable evolution or are there other alternatives? In several papers we can see that also processes, culture, organizational structure need to be focused.

Back et al. [1] illustrate an exemplifying model, which focuses on the topics mentioned above. This approach uses nine dimensions to measure the level of digitalization. A digital readiness score, which is used to see the companies score and a comparison to others, was published by Jahn and Pfeiffer [2]. Digitalization needs to be evolved from a technical perspective to a management one within the business engineering methodology. Business engineering describes the systematic transformation of organizations from the industrial age into the information age in [3]. This approach can also be derived and applied to healthcare providers. The framework of business engineering is appropriate for investigation of transformational effects and helps to gain a holistic view of necessary activities. The difference between classical and digital transformation is the outcome's enrichment by appropriate products or services.

Digital transformation is a conglomeration of different disciplines to realize digital evolution of organizations and can be described as a combination of adaptions in strategic management, business models, organizational structure, process and project management as well as corporate culture by use of digitalization. It is not an evolution that will automatically be done, but rather a gradual change that health care providers should actively force. Furthermore, it is not only a technical topic, but also a topic that concerns the whole company. There are some consulting companies or integrators that still focus on the usage of new information technology to transform health care providers in a digital perspective.

Digital maturity models describe different capabilities in their scopes that are important in transformation. A maturity model was published in [4] by Forrester Research. A more enhanced maturity model containing 8 dimensions and 5 maturity levels was published in [5]. These skills have been subdivided into maturity criteria and the degree to which the organizations meet these abilities.

In addition, the digital maturity models are used to visualize the status quo and to derive future transformation paths for executives and support their decisions. What the models cannot afford is to specify specific transformation instructions. There are no predefined paths for transformation that will follow the same patterns and similarly can be applied to different organizations.

The International Data Corporation (IDC) published a model, called Digital MaturityScape, in which they see digital transformation as multifaceted. This model is illustrated in Fig. 1 and helps organizations to identify the status quo and their future progress in digital transformation. Azhari et al [5] suggested that transforming an organization needs to take place in five maturity dimensions: Leadership, Omni-Experience, Information, Operating Model and WorkSource.

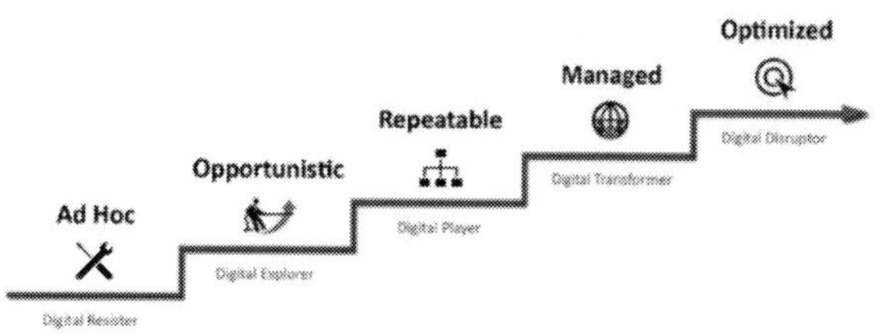

Figure 1. Digital Transformation MaturityScape [5]

IDC already did intensive research and found that 30% of the organizations are at the digital explorer stage, which means that they are working on digital projects. The problem is that the projects are not repeatable and not scalable. Another third are at the digital player stage, which means that they are executing on a repeatable basis, but these digital initiatives are focused in silos. It is terrifying that just 14% of all organizations are at the stage of digital transformers. These numbers arise only from the industrial sector, which can be seen in [6].

Health care providers aspire to solve the following two problems, by use of digital transformation. First, due to limitations in public funding, there is a need to reduce internal costs. Second, to achieve a competitive advantage they have to focus more on their patients. This aspect is just external, but we have recognized that in digital transformation two dimensions seem relevant to the holistic view:

- External: patients
- Internal: employees

In order to achieve excellence in the external dimension, a sustainable internal dimension must exist. Based on the considered maturity models, we have recognized the missing focus on the employees and taken into account in the development of our digital capability model.

III. DIGITAL TRANSFORMATION MODEL

Our digital capability model is derived from different transformation models, such as published in [7], and contains five scopes. The following four are well known:

- Processes
- Organization
- Technology
- Culture

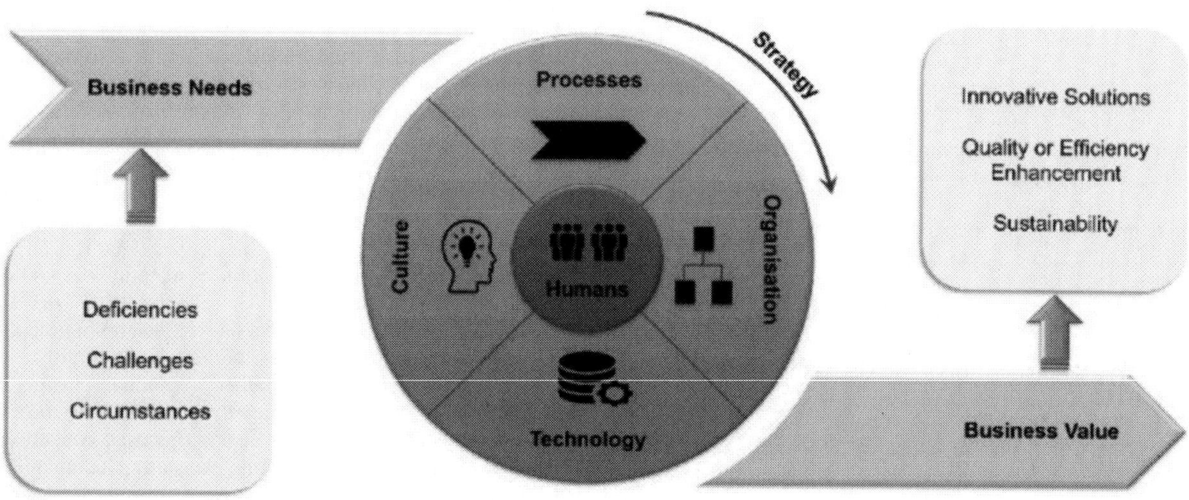

Figure 2. Holistic Transformation Model

The fifth scope is very important in particular, for health care providers: the humans, focused on employees. These human beings have great influence on the other four scopes of our model, which is illustrated in Fig. 2. Their digital intelligence needs to be measured in combination with each of the other scopes to achieve a holistic view. Digital intelligence, in this case, is defined as the ability of humans to naturally deal with technology based systems.

To support decision-making methods it is necessary, to quantify the status quo and the future state of each scope. We have been inspired by radar technology to gain a 360° view and get a valid positioning on a predefined area. Radar is an abbreviation for radio detection and ranging and was patented in 1904. The ideas were based on the perceptions of Heinrich Hertz in 1888, when Hertz detected the polarization dependent reflection of electromagnetic waves, which can be seen in the patent by Christian Hülsmeyer [8].

The idea of Radar can be used for health care providers' transformation paths. We propose the following three steps to derive a digital transformation path:

- Detect
- Position
- Transform

In the first step (detect), organizations have to identify the status quo - the current capability level of processes, organization, technology and culture. In the next step (position), they need to define the degree of capability they want to achieve with their organizations. Finally (transform), the distance between detection and positioning is the transformation path, which has to be realized by the organization. The second and the third step have a massive influence on the decision-making process, because their impact on the transformation paths' activities and complexity is significant.

The boundary condition of each decision needs to be focused, especially depending on the employees' digital intelligence. Therefore, the focus on the humans is important in the decision making process, because it is the key success factor for sustainability. So these steps can be used for decision-support at management level with considerations in the scopes (processes, organization, technology and culture) of our holistic transformation model.

Fig. 3 shows an example of our holistic capability model, called Social & Health Care (SHC) Radar. The current value (shown as a dark grey line) and the target value (shown as a light grey line), which are illustrated as percentage values, are gained for the four scopes (processes, organization, technology and culture). The transformation gap between these two values identifies the transformational needs of organizations. The cultural scope in an exemplary health care provider was currently evaluated with 25%. The organization is planning to reach 45% in that scope. Several methods have to be developed and applied by the organization to eliminate the calculated transformational gap of 20%.

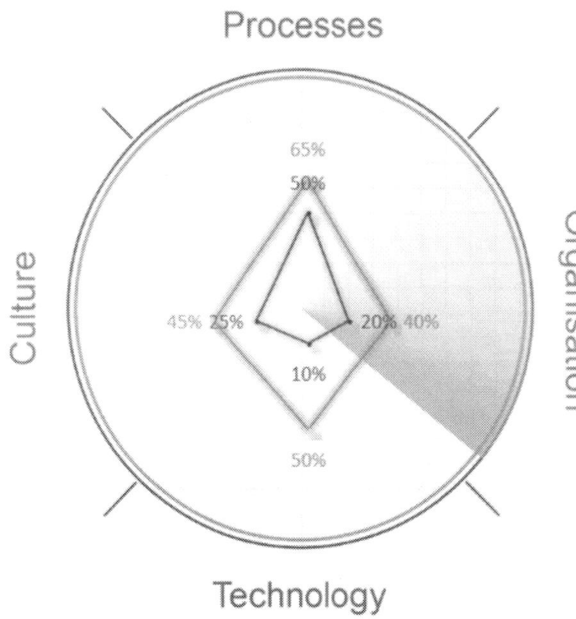

Figure 3. Holistic Capability Model

Table I shows the evaluation of an exemplary health care provider. We recommend to prioritize digital transformation activities based on the amount of the gap.

TABLE I. VALUES OF AN EXEMPLARY HEALTH CARE PROVIDER

Scope of SHC Radar	**Current value**	**Target value**	**Gap**
Processes	50%	65%	15%
Organization	20%	40%	20%
Technology	10%	50%	40%
Culture	25%	45%	20%

IV. CONCLUSION AND FUTURE WORK

Our research gains a valuable insight into different digital maturity models. Consequentially we have recognized the missing focus on the employees and taken this into account in the development of our holistic transformation model. We recommend health care providers to consider their employees' digital intelligence in the decision-making process, because of the strong dependences to all scopes of our model.

Innovative decision-making methods need measureable data. The status quo and future position have to be identified and illustrated in our SHC Radar. In ongoing studies, we define the four scopes more precisely to gain a better level in granularity. Moreover we extend our digital capability model by the fifth dimension: the humans. Our decision-support methodology can be applied for health care providers worldwide to benchmark their digital maturity level.

Furthermore we are going to evaluate whether future technologies such as voice recognition, gesture recognition, artificial intelligence, and others will decrease the need for employees' digital intelligence.

REFERENCES

[1] A. Back and S. Berghaus, "Digital Maturity & Transformation Report 2017", Universität St. Gallen, St. Gallen, 2017

[2] B. Jahn and M. Pfeiffer, "The Digital Revolution - New business models take place (only) new communication", Gabler Verlag, Wiesbaden, 2014.

[3] H. Österle and R. Winter, „Business Engineering", Springer-Verlag, Berlin Heidelberg, 2003.

[4] M. Gill et al, "The Digital Maturity Model 4.0", Forrester Research, Inc., Cambridge, 2016.

[5] P. Azhari, N. Farby, A. Rossmann, B.Steimel and K. Wichmann, "Digital Transformation Maturity Model", neuland & Research Center for Digital Business an der Hochschule Reutlingen, Cologne, 2014.

[6] S. Fitzgerald, "IDC MaturityScape Benchmark: Digital Transformation Worldwide 2017", International Data Corporation, Framingham, 2017.

[7] G. Westerman et al, "Digital Maturity Matrix", MIT Center for Digital Business & Capgemini Consulting, Cambridge, 2012.

[8] C. Hülsmeyer, "To report procedure removed metal objects by means of electric waves a watcher", Imperial Patent Office, Patent Nr. 165546, Berlin, 1904

Chair-Stand Device for the Assessment of Elderly Patients in Risk of Frailty

Alberto Sánchez-Sánchez, Elena Villalba-Mora, Xavier Ferre, Cristian Moral
Centro de Tecnología Biomédica
Universidad Politécnica de Madrid
Madrid, Spain
e-mail: {alberto.sanchez, elena.villalba, xavier.ferre, cristian.moral}@ctb.upm.es

Leocadio Rodriguez-Mañas
Servicio de Geriatría
Hospital de Getafe
Getafe, Spain
e-mail: leocadio.rodriguez@salud.madrid.org

Dieter Hayn
Digital Health Information Systems
AIT Austrian Institute of Technology
Graz, Austria
e-mail: dieter.hayn@ait.ac.at

Raquel Bernabé Espiga, Juan Luis Sánchez Sánchez, Andrea López Diez-Picazo
Fundación para la Investigación Biomédica
Hospital de Getafe
Getafe, Spain
e-mail: {raquel.bernabe, juanluis.sanchez, andreajulia.lopez}@salud.madrid.org

Abstract— **Frailty is an intermediate state in the ageing trajectory, preceding the onset of disability. It can be assessed via clinical, cognitive, nutritional, and physical performance components, whereas one component of physical assessment is measurement of lower limb strength. We propose a device for measuring lower limb strength based on the 30s sit-to-stand test (30s-STS) in an autonomous environment. The device is based on an ultrasound sensor that is mounted on the backrest of a chair and measures the distance to the back of the patient. A signal processing algorithm was developed with a dataset of healthy subjects and evaluated prospectively with geriatric patients. Results of evaluation show that measurement for geriatric patients is poorer than expected. Thus, we plan to re-train the algorithm with additional geriatric patients, and to study further the complexities of the 30s-STS analysing aspects such as the inter-observer-variability of reference annotations. The new algorithm will be evaluated as part of a clinical trial with 40 elderly patients at their dwelling. The clinical trial will also assess the acceptability of such a device in a home setting.**

Keywords- frailty; ultrasound sensor; mHealth.

I. Introduction

Frailty is conceptually defined as a state of older adults with increased vulnerability, resulting from age-associated declines in physiologic reserve and function across multiple organ systems, which compromises the ability to cope with stressors [1][2]. In contrast to permanent disability, frailty advance can be potentially reversed through appropriate interventions [3][4][5].

The project FrAilty Care and wEll funcTion (FACET) targets the development of tools that facilitate the detection of frailty advance, enabling intervention to prevent or delay the onset of frailty. FACET assessment strategy comprises clinical, cognitive, nutritional, functional and physical performance components. Among them, physical performance stands as a strong predictor of undesired clinical outcomes in elderly patients such as deaths, hospitalisation, disability [6], and falls [7]. The measurement of lower limb strength is one of the three basic physical assessment tools along with balance and gait speed assessment [8]. FACET project aims to measure both gait speed [9] and lower limb strength via specialized devices. The latter complements the former, since it directly measures lower limb power, a relevant factor leading to sarcopenia and higher risk of falls.

We present in this paper the device for lower limb strength measurement in the patient's dwelling, to be used together with a tablet, empowering the patients through self-assessment and a continuous and closer follow-up. In the next section the device and the algorithm used to interpret the measurements are described, along with the evaluation results and refinement steps taken accordingly. Section III includes a discussion of such results and, finally, Section IV presents the next steps in this line of research.

II. Methods & Results

The designed device offers the possibility of an automatic and unsupervised 30s-STS [10], counting the times elderly patients stand up and sit down from a chair in thirty seconds. It is attached to the backrest of a chair (see Figure 1), and it is equipped with an ultrasound sensor that continuously measures the distance between the device and the test subject with a sampling frequency of 10 Hz.

It is meant to be used by patients with the help of informal caregivers, to measure lower limb strength according to their geriatrist established plan, (i.e., once or twice a week). A training session is held at patient homes to ensure a safe usage, particularly focusing on avoidance of falling hazards.

A signal pre-processing and measuring algorithm (v1) was developed in MATLAB (The MathWorks, Nattick, MA, USA), and trained with the data from 30s-STS tests with 25 healthy young subjects in a controlled laboratory environment (dataset A). The mean number of stands per test was 11.2 with a standard deviation of 3.9. The pre-processing in v1 consisted of removal of outliers (distance > 70 cm) moving median filtering with a moving window length of 0.7 s detection of local maxima. Each detected peak was interpreted as a successful stand-up-event.

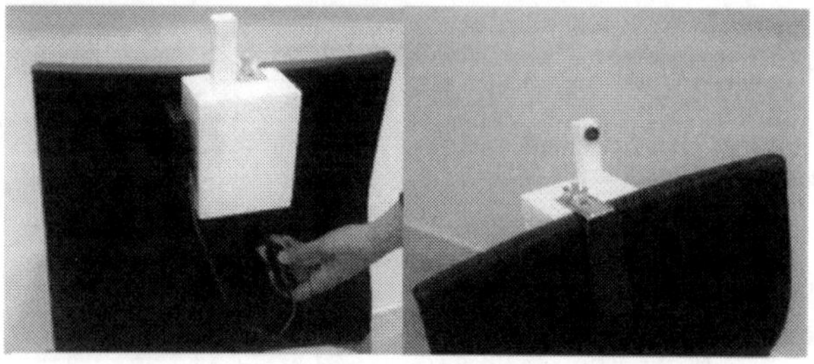

Figure 1. Chair-Stand device as tested with data set B

When applied to dataset A, the first version of the algorithm was able to exactly count 88 % of the measurements (i.e., the calculated number of sit-to-stand events = reference number of sit-to-stand events as annotated via human counting). The mean absolute error was 0.16 stands and the mean relative error was 2.22 %.

This v1 of the algorithm was evaluated in a pre-clinical environment at the University Hospital of Getafe (HUG) with 30 volunteer geriatric patients and a trained nurse manually counting the stands in 30 s (dataset B). The mean number of stands was 9.2 ± 1.7 stands per test. The results of this prospective test showed a dramatically lower accuracy of the device in the assessment of potential real subjects. Only 7 % of measures were consistent with the manual measurements by the trained nurse. The mean absolute error was 2.57 stands and the mean relative error was 27.49%.

Based on dataset B, the algorithm was refined to a v2, trained retrospectively with dataset B. An adaptive threshold was defined based on the mean value of the moving median and minimum within a 4s window. The threshold for outlier detection was increased from 70 to 99cm. Successful stand-up-events were counted whenever a) the moving minimum of the signal with a window length of 0.7 s exceeded this threshold and b) the moving median within a 0.7 s window right after this timepoint exceeded 25 cm. Events that were closer than 1s to one another were not considered. V2 of the algorithm improved the accuracy of the measurements up to 40%. The mean absolute error was 1.10 stands and the mean relative error was 12.80 %.

III. Discussion

We found that the results achieved with the v1 of our algorithm were rather poor when applied to elderly patients. Due to the heterogeneous biomechanics of elderly patients while performing the test, the data differed remarkably as compared to healthy subjects. This fact, along with the poor adaptability of the algorithm to different chairs and subjects, were identified as potential sources of error in the measurement. V2 of the algorithm offered better results, but overfitting cannot be excluded.

In some cases, it is hard to tell whether a sit-to-stand event should be counted or not, i.e., whether the patient reached a complete upright position. Especially, the decision whether the final event was within the 30 s time interval is sometimes hard to make. Therefore, even our reference annotations (manual counts of sit to stand events) might show some uncertainties. Considering the study itself, we can identify as limitations the sample size and the current status of the device as a prototype.

For the system to be used in a clinical setting, it will be necessary to improve the accuracy by feeding the algorithm with additional measurements of real patients. The aim will be to make it at least as good as the average manual measurement by health professionals, which will also need to be measured. Nevertheless, for the system to be useful for monitoring frailty in a home setting, it is only necessary that the measurements are consistent within the same subject, to ensure decline in lower limb strength is detected early.

IV. Further Steps

In a next step, our algorithm will be evaluated in a prospective way with an additional test-set, which will be recorded from patients at the HUG. During this next step, it is planned that more than one person will manually count the sit-to-stand events, in order to quantify the inter-observer-variability of our reference annotations.

Next, the device will be used in a clinical trial with 40 elderly patients at their dwelling. The collected data will be analysed, inquiring the clinical relevance of a home-based chair stand test, and its ability to prevent frailty evolution to disability.

Acknowledgments

This research was funded by the EIT Health in the FACET (FrAilty Care and wEll funcTion) project.

References

[1] L. P. Fried, et al., "Frailty in older adults: evidence for a phenotype," J Gerontol A Biol Sci Med Sci, vol. 56(3), pp. M146-157, 2001.

[2] X. Chen, G. Mao, and S. X. Leng, "Frailty syndrome: an overview," Clinical Interventions in Aging, vol. 2014:9, pp. 433-441, 2014, doi:10.2147/CIA.S45300.

[3] L. Rodriguez-Mañas, and L.P. Fried, "Frailty in the clinical scenario," Lancet Lond Engl, vol. 385, pp. e7–e9, 2015.

[4] L.R. Pollack, et al., "Patterns and Predictors of Frailty Transitions in Older Men: The Osteoporotic Fractures, " Men Study. J Am Geriatr Soc, vol. 65, pp. 2473–2479, 2017.

[5] M. Cesari, et al., " A physical activiy intervention to treat the frailty syndrome in older persons – results from the LIFE-P study," J Gerontol Ser A Biol Sci Med Sci, vol. 70(2), pp. 216–222, 2014.

[6] D. Legrand, et al., "Muscle Strength and Physical Performance as Predictors of Mortality, Hospitalization, and Disability in the Oldest, " Journal Of The American Geriatrics Society, vol. 62(6), pp. 1030-1038, 2014, doi: 10.1111/jgs.12840.

[7] A. Wihlborg, M. Englund, K. Åkesson, and P. Gerdhem, "Fracture predictive ability of physical performance tests and history of falls in elderly women: a 10-year prospective study," Osteoporosis International, vol. 26(8), pp. 2101-2109, 2015, doi: 10.1007/s00198-015-3106-1.

[8] J.M. Guralnik, et al., "A short physical performance battery assessing lower extremity function: association with self-reported disability and prediction of mortality and nursing home admission," J Gerontol, vol 49, pp. M85-94, 1994.

[9] X. Ferre, et al. "Gait Speed Measurement for Elderly Patients with Risk of Frailty". Mobile Information Systems. Hindawi, vol. 2017, 2017.

[10] C. J. Jones, R. E. Rikli, W. C. Beam, "A 30-s chair-stand test as a measure of lower body strength in community-residing older adults," Research Quarterly for Exercise and Sport, vol. 70, pp. 113-119, 1999.0

Prediction of Patient Outcomes after Renal Replacement Therapy in Intensive Care

Harry F. da Cruz*, Siegfried Horschig*, Christian Nusshag**, and Matthieu-P. Schapranow*
* Hasso Plattner Institute, Digital Health Center, Rudolf-Breitscheid-Str. 187, 14482 Potsdam, Germany
** Heidelberg University Hospital, Department of Nephrology, Im Neuenheimer Feld 162, 69120 Heidelberg, Germany

E-Mail: *{harry.freitasdacruz|schapranow}@hpi.de, siegfried.horschig@student.hpi.de,
**christian.nusshag@med.uni-heidelberg.de

***Abstract*—In order to compensate severe impairments of renal function, artificial, extracorporeal devices have been developed to enable Renal Replacement Therapy. The parameters utilized for this procedure and the specific patient characteristics substantially affect individual patient outcomes and overall disease courses. In this paper, we present a clinical prediction model for outcomes of critically ill patients who underwent a specific form of renal replacement, hemodialysis. For this purpose, we employed two machine-learning models: Bayesian Rule Lists and Multi-Layer Perceptron. To provide more transparency to the perceptron model, we applied mimic learning to its output based on a Bayesian Ridge Regression model. Results show that while the perceptron model outperforms the rule-based classifier, the use of the mimic learning approach enables more thorough model scrutiny by a medical expert, revealing possible model biases, which might have gone unnoticed, a sensitive issue in a high-stakes domain such as medicine.**

***Keywords*-Clinical Prediction Model; Renal Replacement Therapy; Machine Learning; Supervised Learning.**

I. Introduction

The renal system in the human body has the purpose to excrete predominantly water-soluble metabolites and toxins in order to maintain a sufficient blood homoeostasis [1]. If this system is impaired severely, e.g., in the context of an Acute Kidney Injury (AKI), artificial, extracorporeal organ replacement therapy becomes necessary [2]. Therefore, different Renal Replacement Therapy (RRT) modalities are available. One example is the hemodialysis, where the solute exchange takes place via diffusion across a semipermeable membrane between the blood and the dialysate or dialysis fluid [3].

Dialysis outcomes are highly dependent on both the patient's characteristics and clinical parameters, as well as on the type of the RRT procedure applied [4]. Furthermore, RRT modalities based on a filtration circuit, such as hemofiltration or hemodiafiltration are particularly costly, requiring specialized equipment and nursing staff [5]. In addition, various parameters have to be adjusted for each patient, e.g., duration of the process, the filtration rate and flow rates of the blood and dialysate. Clinical prediction models can aid in decision making by providing nephrologists with more accurate prognostic information under uncertainty of outcomes [6].

Aside from usual criteria like accuracy or recall, when employing a machine learning model in the medical context, one especially important factor is the interpretability of the model, since doctors must take full responsibility for the respective decision and therefore require a high degree of trust on the model [7]. As such, one can roughly distinguish between o categories of machine learning algorithms: interpretable and non-interpretable. One example for interpretable models are Bayesian Rule Lists (BRL) [8]. By presenting itself as *if...then...else* lists, it is easy for humans to comprehend both the decision making and the individual influence of each parameter on the outcome. In contrast, the Multi-Layer Perceptron (MLP) model is usually more accurate, but non-interpretable, since the weights of the nodes in the hidden layers are all that is exposed to the outside. Due to the fact that different loss and activation functions take effect when updating those weights, the abstraction to the original input data is just too cumbersome for a human to grasp.

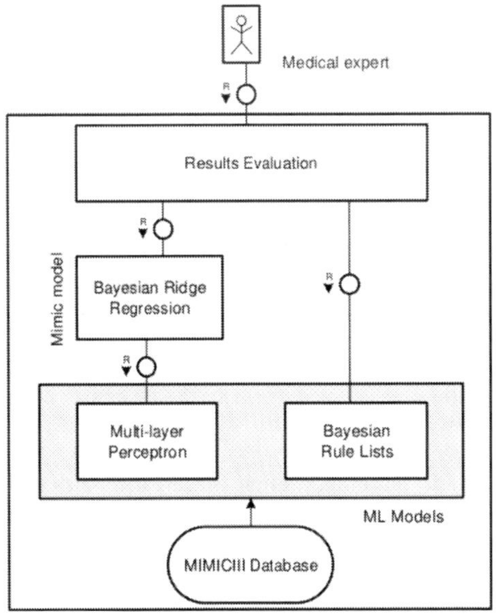

Figure 1. Our machine-learning setup modeled as a FMC block diagram. We incorporated a multi-layer perceptron and bayesian rule lists machine-learning model.

In order to overcome the tradeoff between interpretability and accuracy, we employed a strategy called mimic learning [9]. By training an interpretable model on the predictions of the more accurate, non-interpretable model, we gain insight into its decision process and can therefore enhance the non-interpretable model's intelligibility.

Our contribution consists of a Clinical Prediction Model (CPM) to prognosticate patient-specific outcomes after RRT in the Intensive Care Unit (ICU) as modeled in Figure 1 using Fundamental Modeling Concepts (FMC) block diagram. We evaluated the performance of two different models, BRL as the interpretable variant and MLP as its non-interpretable counterpart. After that, we employed mimic learning to help overcome the tradeoff between accuracy and interpretability and provide some insight into the decision parameters of the MLP. We then interviewed an expert in the field of Nephrology for scrutiny of the models thus developed.

The remainder of the work is structured as follows: In Section II we place our work in the context of related work. We present our incorporated data and models in Section III and present results of our work in Section IV. We discuss our findings in Section V followed by the conclusion in Section VI.

II. Related Work

Machine learning research in Nephrology has been traditionally geared towards kidney disease detection using decision trees and naïve Bayes [10, 11]. However, those models tend to be less accurate when compared to more advanced models, which prompted the community to experiment with other methods. Vijayarani and Dhayanand and Sinha and Sinha used Support Vector Machine (SVM) and Artificial Neural Network (ANN) for prediction of kidney disease with encouraging results [12, 13]. In a similar fashion, Lakshmi et al. compared the three models regression, random forest and ANN, proposing the latter for better performance and accuracy [14].

The enhanced performance with modern machine learning tools, however, is achieved at the expense of model interpretability. The ability to explain and interpret decision is a key requirement in medical applications. In the context of machine learning, Lipton placed the particular focus was on identifying decision boundaries and ascertaining the influence of specific feature for improved interpretability [15]. Approaches have been developed to achieve interpretability of black box models, such as the classification vectors approach by Baehrens et al. and the Locally-Interpretable Model-agnostic Explanations (LIME) by Ribeiro et al. [16, 17]. In particular, Katuwal and Chen applied the LIME technique for achieving interpretability of random forests for predicting ICU mortality, achieving accuracies of 80 % [7]. Still in the medical domain, Hayn et al. quantified the influence of individual features on particular decisions made by a random forest in clinical modeling applications [18].

Unlike previous work, we focus specifically on the task of outcome prediction of RRT patients while comparing two types of models side-by-side, one interpretable (BRL) and another non-interpretable (MLP). For aiding the interpretability of the complex model, we made use of the mimic learning technique as proposed by Che et al. in lieu of the LIME method employed in extant research, because we aim to obtain a global understanding of the model's inner workings rather than explain individual instances of classification [7, 9]. Che et al. used Gradient Boosting Trees as mimic learning model while we applied Bayesian Ridge Regression (BRR) since their output more closely resembles logistic regression, a technique widely employed in medicine.

III. Methods

In the following, we share details about methods and data employed for our clinical models.

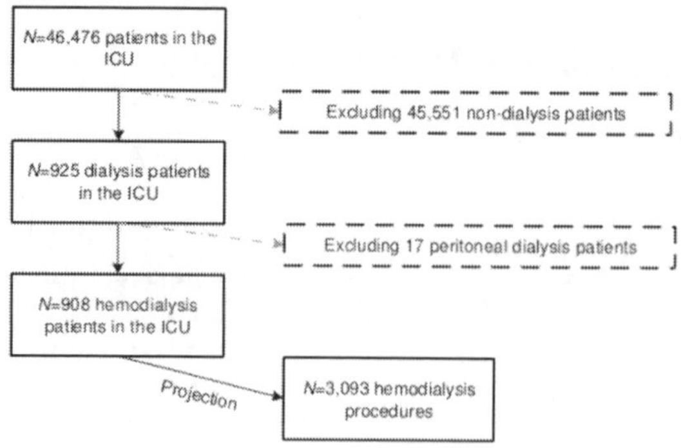

Figure 2. Cohort selection of the dialysis procedures based on the MIMIC III intensive care patients.

A. Tools

We used *RapidMiner* [19], which allowed us to prepare data, develop and cross-validate first models. The final models were subsequently implemented with the *scikit-learn* library [20] in Python 2.7. The data we used were provided by the MIMIC III dataset [21] stored in an in-memory database via an Open Database Connectivity interface [22].

B. Data

The *MIMIC III* dataset contained hospital admission data for patient collected over an eleven-year period in a Boston hospital. As seen in Figure 2, out of the approximately 46,000 patients present in the dataset, we extracted 908 relevant patients for this paper, totaling approximately 3,000 dialysis procedures for model training. We had to exclude from the analysis patients who had undergone peritoneal dialysis, since they are not relevant in an acute context.

The cohort does not contain patients who underwent hemofiltration or hemodiafiltration, only hemodialysis patients. Under hemodialysis, the data comprises both Continuous Renal Replacement Therapy (CRRT) and Intermittent Hemodialysis (IHD) modalities, therefore RRT type was a feature in the final model. We therefore derived another cohort only with CRRT patients (N=1,163 procedures) and IHD patients (N=1,930 procedures) to ascertain whether results were consistent across dialysis modalities. We further derived a cohort consisting exclusively of acute patients (N=954 procedures) since patients, who presented acute kidney injury without previous history of renal disease, present peculiarities from a clinical standpoint.

In cooperation with the *Nierenzentrum Heidelberg*, we conducted interviews with a subject-matter expert in order to curate a list of suitable features, amounting to about 80 predictors. Those included patient demographics, such as age or Body Mass Index (BMI), RRT parameters such as the duration of the procedure, comorbidities as well as lab values, including parameters such as serum creatinine and Glomerular Filtration Rate (GFR) for 24, 48 and 72 hours before the procedure. Additionally, we included patient vitals and outcomes such as 90-day mortality, renal recovery, mechanical ventilation days and length of stay in the ICU.

Missing Data: Due to the manually curated nature of the MIMIC III dataset, aside from occasional data inconsistencies, a significant amount of data was missing. For example, the columns containing serum creatinine and GFR values before the procedure were missing in approx. 20 % of samples. As the scikit-learn models need a complete dataset for training, we decided to impute the missing values using k-nearest neighbors algorithm (k-NN) [23].

C. Models

In the following, we describe the models and strategies used as well as the parameters chosen for training for both the interpretable and non-interpretable algorithms, as well as the interpretability approach employed.

1) Interpretable - Bayesian Rule Lists: For the interpretable model, we chose the existing Python 2 implementation of BRL [8]. Letham et al. describe it as a direct competitor to decision tree approaches, as the model achieves high accuracy for classification tasks while still being intelligible for subject-matter experts. This algorithm tries to derive *if...then...else* statements over a dataset with the important criteria of their being sparse for better human readability. It builds Bayesian association rules consisting of an antecedent a and a consequent b. The consequent has a multinomial distribution over all the predicted labels y, so that the rules are defined in Equation 1.

$$a \rightarrow y \sim Multinomial(\theta) \tag{1}$$

Mining antecedents from the data generated these rules and afterwards computing the posterior consequent distribution over the antecedent lists. BRL have the advantage of being easy to interpret due to their sparsity while retaining accuracy in classification. However, there are algorithms providing a higher accuracy, which also have the capability of more elaborate parameter tuning. Additionally, the current implementation of BRL has the shortcoming of a very long runtime and only being able to classify binary targets. Thus, we had to adjust the target features accordingly through use of binary operator for continuous predictors.

Parameters: The sole adjustable parameter in the implementation used was the maximum number of iterations. Multiple adjustments to this parameter – incl. changes by a factor of ten – did not result in a significant change, neither for the runtime nor for the accuracy. For the evaluation, we chose a value of 50,000 maximum iterations.

2) Non-Interpretable: Deep Neural Network: As non-interpretable model, we chose the scikit-learn implementation of MLP, which is able to handle both regression and classification tasks. Just as other implementations, this network consists of multiple layers of so-called "neurons": one input layer with as many neurons as there are inputs, one output layer with the size of the number of target features and hidden layers varying in size and quantity. The log-loss function is optimized through updating weights for each neuron for each iteration of model training. The neural network can be defined as mathematical function $f(x)$ as shown in Equation 2 with the activation function K and k-times $g_i(x)$ representing the dependencies between functions with an individual weight w_i.

$$f(x) = K\left(\sum_{i=1}^{k} w_i g_i(x)\right) \tag{2}$$

MLP are a widely used form of machine learning due to their versatility and high accuracy. They provide a wealth of parameters to tune, but finding the right ones for a specific use case can prove cumbersome. Furthermore, the decision making process of such a neural network is not comprehensible to a human and thus provides nearly no interpretability.

Parameters: The amount of parameters to be adjusted when using neural networks is very high. Performing grid search over some of the parameters, we found the default ones from the library to perform the best.
This means the learning rate, which determines the speed and accuracy of convergence, was set to 0.001. The activation function, determining the output of the neurons in the hidden layer, was the rectifier linear unit "relu". The network consisted of one hidden layer with 100 neurons. We set the maximum number of iterations before convergence was set to 200.

3) Interpretability Approach: Mimic Learning: The large amount of neurons in the MLP and the many parameters influencing their weights and output make it very difficult – if not impossible – for a human to understand the influence of each feature on the training. Therefore, we aimed to provide some insight into the workings of the MLP by applying a method called mimic learning. Building upon the approach of Che et al. we trained an interpretable model – the so-called mimic model – on the outputs of the non-interpretable model (MLP). In this approach, the mimic model takes on the same input features as the non-interpretable model.

For classification tasks, the outputs of the non-interpretable model are termed soft scores, because as they are probabilities, they are continuous variables, coming close to the actual target features. Training the mimic model on the soft scores allows creating a much smaller, thus understandable, faster but still equally accurate model. Using the principle Che et al. called knowledge distillation, it is even possible for the mimic model to generalize better than the non-interpretable model [9]. This happens because the non-interpretable model filters out certain noise in the training data, which could have a negative impact on training performance of the interpretable model. For the mimic model, we needed an algorithm, which was able to predict continuous scores in order to train it on the aforementioned soft scores. For this purpose, we utilized Bayesian Ridge Regression.

Bayesian Ridge Regression: Similar to common linear regression, this algorithm tries to find coefficients for each input feature so that they map to the target feature, minimizing loss. In addition to common linear regression, it includes regularization parameters to control the growth of the coefficients. Therefore, this model is less prone to over-fit while still being as fast as linear regression.

Furthermore, regression in general has the advantage of being very fast concerning training time and interpretable, as one can easily inspect the coefficients for each feature. However, due to the simplicity of regression models, they usually lack accuracy when compared to more elaborate algorithms. Very few parameters can be adjusted for this algorithm and for our experiments, we applied the default ones. This means that all regularization parameters were set to 10^{-6} and the number of iterations before convergence were set to 300.

The process logic implemented for the mimic learning approach is shown in pseudo-code in Algorithm 1.

Algorithm 1: Mimic Learning with BRR

Input: MLP model, Training dataset and Test dataset
Result: Sorted mimic regression coefficients
Obtain soft scores from MPL on Training dataset;
Train BRR model on soft scores and Training dataset;
Apply trained BRR model on Test dataset;
Obtain BRR regression coefficients on Test dataset;
Sort regression coefficients;
Return regression coefficients;

IV. RESULTS

In the following section, we compare the performance of our interpretable model, the BRL, and our non-interpretable model, the MLP. Although there were continuous values for our target variables in the dataset, we had to transform them into a binary format in order for the BRL classifier to work. Therefore, we considered the following outcomes:

- **90-days Mortality:** Indicates whether the patient has died within a 90-day period (1 = dead / 0 = alive),
- **Renal Recovery:** If patient has been for more than 7 days without dialysis requirement, renal function is considered to be restored (1 = recovery / 0 = no recovery),
- **Ventilation Days:** Indicates whether the patient has been on ventilation for been less than seven days (1 = true / 0 = false), and
- **Length of Stay:** Points out if length of stay has been less than 7 days (1 = true / 0 = false).

The complete list of features can be found in Table A.I.

Table I shows general performance of the employed classifiers according to the AUCROC performance metric. As expected, the MLP outperforms the BRL classifier in for virtually every patient cohort and patient outcomes, excepting the prediction for ventilation days. The mimic approach using BRL trailed right along the MLP, presenting similar results. Concerning runtimes, there were considerable differences between the two classifiers. While the MLP took only a few seconds to conduct the full training with the configuration described previously, the BRL needed up to one hour to train on the same data. Due to the interpretable nature of the BRL, a medical expert can analyze the importance of single features directly on the model output.

Figure 3 shows the influence of some features and their values on the prediction of 90-day mortality. For this outcome, the Sequential Organ Failure Assessment (SOFA) score was a key feature. This score is widely used in intensive care for this very purpose, therefore the BRL classifier correctly detected this. "CR_24_B" corresponds to blood creatinine 24h before the hemodialysis procedure and Elixauser is a comorbidity score. High values for both of these features are associated with increased mortality, but from the output of the BRL alone it is hard to ascertain whether it correctly captured this relationship.

```
IF SOFA: 0.69_to_inf THEN probability of DIED_90DAYS:
80.3% (73.1%-86.6%)

ELSE IF CR_24_B: 0.153_to_inf AND ELIXHAUSER: -inf_to_0.31
THEN probability of DIED_90DAYS: 3.0% (1.0%-6.1%)

ELSE IF LACTATE: 0.015_to_0.056 AND CR_72_B: -inf_to_0.18
THEN probability of DIED_90DAYS: 35.4% (29.5%-41.6%)

ELSE...
```

Figure 3. Excerpt of the rules from the Bayesian Rule Lists classifier when predicting 90-day mortality. Abbreviations: SOFA = Sequential Organ Failure Assessment score, CR_24_B, CR_72_B = Serum Creatinine 24h and 72h before procedure, respectively.

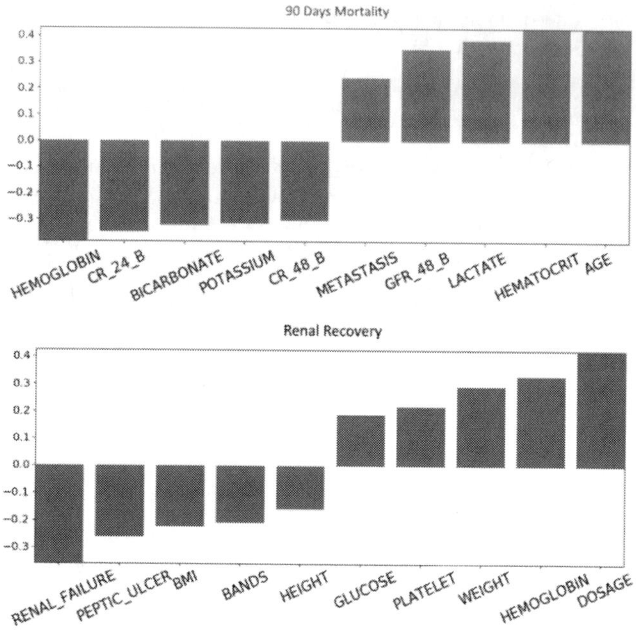

Figure 4. Coefficients of the most important features (last five and top five after sorting) for the Bayesian Ridge Regression trained as mimic model for 90-days mortality and renal recovery. Abbreviations: CR_24_B, CR_48_B, GFR_48_B = Serum Creatinine and Glomerular Filtration Rate 24h and 48h before procedure, respectively, BMI = Body-Mass Index.

For the MLP results to be inspected, we had to apply the mimic learning strategy discussed. First, we needed to evaluate if the performance of the mimic model is satisfactory when being trained on the outputs (soft scores) of the MLP. One can verify that, while the BRR is still worse than the MLP, it performed better than the BRL, if only by a small margin. It is important to highlight, however, that the mimic classifier is only as good as the predictor it originally learned from.

In Figure 4, we can assess the influence of single features on a positive prediction of both 90-day mortality and recovery of renal function. For example, the higher the rightmost feature, e.g., the age of the patient, the higher is the probability of the patient to die within 90 days. Conversely, the higher the leftmost feature, e.g., the hemoglobin value in the blood of the patient, the less likely the patient is to die within 90 days. These results were submitted to the appraisal of a Nephrology expert to establish clinical relevance and adequacy.

Outcome	Complete cohort			Acute patients			IHD patients			CRRT patients		
	MLP	BRL	BRR	MLP	BRL	BRR	MLP	BRL	BRR	MLP	BRL	BRR
90-days mortality	0.84	0.76	0.79	0.83	0.79	0.81	0.83	0.74	0.79	0.77	0.72	0.72
Recovery of renal function	0.91	0.88	0.88	0.86	0.68	0.79	0.90	0.87	0.91	0.86	0.79	0.84
Ventilation days <7	0.81	0.75	0.80	0.64	0.68	0.65	0.81	0.78	0.79	0.77	0.79	0.79
ICU stay days <7	0.83	0.82	0.82	0.78	0.69	0.73	0.80	0.78	0.80	0.73	0.73	0.73

TABLE I. Simulation results displaying Area Under the Receiver Operating Characteristic Curve (AUCROC) for the different analysis cohorts and patient outcomes. Abbreviations: IHD = Intermittent Hemodialysis, CRRT = Continuous Renal Replacement Therapy, MLP = Multi-Layer Perceptron, BRL = Bayesian Rule Lists and BRR = Bayesian Ridge Regression.

V. Evaluation and Discussion

From a classification performance standpoint, our performed experiments suggest MLP as a suitable classifier for the given tasks, with BRL as a close second. MLP performed particularly well for renal recovery prediction, a key outcome for nephrologists. However, both approaches have issues that may hinder adoption in clinical practice.

For example, some of the features deemed important for MLP make sense from a medical standpoint, such as higher age correlating with a higher chance of mortality. However, the results also indicate that high levels of creatinine are associated with lower mortality, which contradict observations in clinical practice. Additionally, as per Figure 4 Glomerular Filtration Rate (GFR), a measure of how well the kidneys are functioning, is associated with higher mortality, a likewise counterintuitive outcome.

Similarly for renal recovery prediction, where high weight and glucose levels are associated with poor outcomes, what contradicts expert knowledge. Furthermore, usually there are non-linear correlations between certain blood values and mortality (e.g., U-shaped curve), such as potassium, as either too low or too large values can influence the patient's health negatively. Such relationships cannot be adequately represented by the mimic learning approach utilized.

It is important to note that these potential spurious correlations are only illuminated through model interpretability, be it because of the nature of the model or the application of mimic learning. Thus, the model interpretability approach employed gives us the possibility to examine the correlations and create assumptions which otherwise might just go unnoticed when using non-interpretable models. The same observations apply for the output of the BRL. For instance, higher lactate values usually lead to other complications, but the upper bound of "infinity" is not meaningful in clinical practice. In order to refine and validate those assumptions, it is necessary to go further with the data analysis. Finding actual upper and lower bounds in the dataset can provide some insight to the actual values the model considers when making predictions.

Additionally, missing data may have a significant influence on the quality of the predictions and certain features could be dropped if they are missing a large amount of values. By training the regression as a mimic model, we can make assumptions on how the MLP *may* make its decisions. There still is a gap between the performance of the regression model and the MLP, which makes it difficult to say how close those coefficients are to the actual influence of features in the MLP. The mimic model performs worse when being trained on the outputs of the MLP as opposed to being trained on the real targets, because it most probably also learns the errors of the MLP. This can be a resolvable issue by improving the performance of the MLP through further parameter tuning and data preparation.

VI. Conclusion

In this paper, we compared the performance of different models when being used in the prediction in the renal context. An important part is the interpretability of such models to validate their applicability for decision support. We used a mimic learning approach to make a MLP interpretable and compared this output to that of the interpretable model, the BRL. Preliminary results for prediction of 90-day mortality enable the exploration of interpretability, showing the influence of single features.

Future work includes more elaborate use of the data, meaning inclusion of more features, more elaborate imputation strategy and collection of more information about the patients. The binary classification limitation of the current implementation of BRL can be overcome by using a more advanced algorithm, such as the one proposed by Yang et al. [24]. When a higher precision is achieved, interpretable models can be used in the context of a clinical decision support system, allowing the doctors to validate the decisions and giving patients insight into their treatment. Likewise, deployment in a clinical setting requires external validation using datasets from different institutions. Subsequently, an impact analysis of the use of such models in a clinical setting should be conducted to ascertain the impacts on care delivery and patient outcomes.

Acknowledgments

Parts of the given work were generously supported by a grant of the German Federal Ministry of Economic Affairs and Energy (01MD15005C).

References

[1] E. Marieb and S. M. Keller, *Essentials of Human Anatomy & Physiology*. Pearson, 2018.

[2] C. Ronco *et al.*, "Renal Replacement Therapy in Acute Kidney Injury: Controversy and Consensus," *Critical Care*, vol. 19, no. 1, pp. 1–11, 2015.

[3] G. M. Fleming, "Renal Replacement Therapy Review," *Organogenesis*, vol. 7, no. 1, pp. 2–12, 2011.

[4] S. Negi *et al.*, "Renal Replacement Therapy for Acute Kidney Injury," *Renal Replacement Therapy*, vol. 2, no. 1, p. 31, 2016.

[5] L. Forni and P. Hilton, "Continuous Hemofiltration in the Treatment of Acute Renal Failure," *New England Journal of Medicine*, vol. 336, no. 18, pp. 1303–1309, 1997.

[6] L. M. Cohen *et al.*, "Predicting Six-month Mortality for Patients Who Are on Maintenance Hemodialysis," *Clin J Am Soc Nephrol*, vol. 5, no. 1, pp. 72–79, Jan 2010.

[7] G. J. Katuwal and R. Chen, "Machine Learning Model Interpretability for Precision Medicine," https://arxiv.org/abs/1610.09045 [retrieved: August, 2018], Oct 2016.

[8] B. Letham *et al.*, "Interpretable Classifiers Using Rules and Bayesian Analysis: Building a Better Stroke Prediction Model," *Annals of Applied Statistics*, vol. 9, no. 3, pp. 1350–1371, 2015.

[9] Z. Che *et al.*, "Interpretable Deep Models for ICU Outcome Prediction," in *Proceedings of the AMIA Annual Symposium*, vol. 2, 2016, pp. 371–380.

[10] P. S. Baby and P. Vital, "Statistical Analysis and Predicting Kidney Diseases using Machine Learning Algorithms," *J of Eng Res & Tech*, vol. 4, no. 07, pp. 206–210, 2015.

[11] R. Greco *et al.*, "Decisional Trees in Renal Transplant Follow-up," *Transplant Proc*, vol. 42, no. 4, pp. 1134–1136, may 2010.

[12] S. Vijayarani and S. Dhayanand, "Kidney Disease Prediction Using SVM and ANN," *Comp Bus Res*, vol. 6, no. 2, pp. 6–17, 2015.

[13] P. Sinha and P. Sinha, "Comparative Study of Chronic Kidney Disease Prediction using KNN and SVM," *J of Eng Res & Tech*, vol. 4, no. 12, pp. 608–612, 2015.

[14] K. R. Lakshmi, Y. Nagesh, and M. Veerakrishna, "Performance Comparison of Three Data Mining Techniques for Predicting Kidney Dialysis Survivability," *J of Adv in Eng & Tech*, vol. 7, no. 1, pp. 242–254, 2014.

[15] Z. C. Lipton, "The Mythos of Model Interpretability," in *Proceedings of the Workshop on Human Interpretability in Machine Learning*, 2016, pp. 96–100.

[16] D. Baehrens *et al.*, "How to Explain Individual Classification Decisions," *Mach Learning Res*, vol. 11, pp. 1803–1831, 2010.

[17] M. T. Ribeiro, S. Singh, and C. Guestrin, "Why Should I Trust You? Explaining the Predictions of Any Classifier," in *Proceedings of the Workshop on Human Interpretability in Machine Learning*, 2016, pp. 91–95.

[18] D. Hayn *et al.*, "Plausibility of Individual Decisions from Random Forests in Clinical Predictive Modelling Applications," *Studies in Health Technoly and Informatics*, pp. 328–335, 2017.

[19] M. Hofmann and R. Klinkenberg, *RapidMiner: Data Mining Use Cases and Business Analytics Applications*. CRC Press, 2013.

[20] F. Pedregosa *et al.*, "Scikit-learn: Machine Learning in Python," *J of Machine Learning Res*, vol. 12, pp. 2825–2830, 2011.

[21] A. E. W. Johnson *et al.*, "MIMIC-III, a Freely Accessible Critical Care Database," *Scientific Data*, vol. 3, 2016.

[22] F. Färber *et al.*, "SAP HANA database," *ACM SIGMOD Record*, vol. 40, no. 4, pp. 45–51, 2012.

[23] J. M. Keller, M. R. Gray, and J. A. Givens, "A fuzzy K-nearest Neighbor Algorithm," *IEEE Transactions on Systems, Man, and Cybernetics*, vol. SMC-15, no. 4, pp. 580–585, jul 1985.

[24] H. Yang, C. Rudin, and M. Seltzer, "Scalable Bayesian Rule Lists," https://arxiv.org/pdf/1602.08610.pdf [retrieved: August, 2018], Feb. 2016.

Appendix

In Table A.I, we share the complete list of features used for our models.

Category	Feature
Demographics	Age Height, Weight, BMI Ethnicity Gender
Dialysis-related	Dosage Modality AKI stage
Comorbidities	Aids Alcohol abuse Blood loss anemia Cardiac arrhythmias Chronic pulmonary Coagulopathy Congestive heart failure Deficiency anemias Depression Diabetes complicated, Diabetes uncomplicated Drug abuse Elixhauser Vanwalraven Fluid electrolyte imbalance Hypertension Hypothyroidism Liver disease Lymphoma, Metastatic cancer, Solid tumor Obesity Other neurological disorders Paralysis Peptic ulcer Peripheral vascular Psychoses Pulmonary circulation Renal failure Rheumatoid arthritis Valvular disease Weight loss
ICU scores	OASIS SOFA SOFA Renal SAPS
Vitals	Heartrate Systolic Blood pressure Diastolic Blood pressure Mean Blood pressure Respiratory Rate Temperature °C Oxigen Saturation (SpO_2)
Laboratory values	Aniongap Albumin Bands Bicarbonate Bilirubin Blood Urea Nitrogen Creatinine 24, 48 and 72h before procedure Chloride Glucose Hematocrit Hemoglobin Lactate Platelet Potassium PTT, INR, PT Sodium WBC Glomerular Filtration Rate 24, 48 and 72h before procedure

TABLE A.I. Model features. Note that related features are grouped together.

Towards a Clinical Risk Reduction: A Hospital Information System Based on Software and Hardware Integration

Raffaele Chianese
Department of Healthcare Solutions
KIRANET I.C.T Research Center
Aversa, Italy
e-mail: Raffaele.chianese@kiranet.it

Francesco Gargiulo
Healthcare Services
Napoli, Italy
e-mail: francescogargiulo@inwind.it

***Abstract*— This article presents FID-MED, an integrated platform composed of a web based clinical application, named FIDCARE, and a medical cart called FIDRUN and describes the possibilities that FID-MED offers, both for quality assurance and risk management, and for cost reduction. FID-MED is an industry research product and the design effort aims to reduce the risk in healthcare services by means of a deep integration between hardware and software and leads to a Hospital Information Systems (HIS) including a specifically designed medical cart. The introduction of key concepts, as abstract drug, and the overall and complete control of clinical processes provides valid tools for the control and rationalization of costs. FID-MED aims to overcome barriers that prevent the adoption of a HIS using enabling technologies – Radio-Frequency IDentification (RFID) and Internet of Things (IoT) – adherence to the main international standards – Health Level 7 (HL7), Digital Imaging and COmmunications in Medicine (DICOM®), etc. – and the ease of integration with third-party software.**

Keywords-Hospital Information System (HIS), FID-MED, FIDCARE, FIDRUN, healthcare risk management, healthcare cost control, medical cart.

I. Introduction

In the Health sector, risk management and cost control are very important. Risk management means the adoption of a set of actions and measures aiming to improve the quality of health services and to ensure the safety of the patient. The control of cost is related to the best management and scheduling of the available resource, such as expensive materials, products and drugs. For these reasons, the possibility to keep under control the involved processes, i.e., medical prescriptions, administrations of drugs, stock management and supply, etc., can lead to a reduction of the possibility of error and economic losses and to savings in time.

Errors in drug therapy represent an important percentage of medical errors and outline not only a clinical problem, with regard to the quality of care, but also an economic problem, with regard to the major cost derived [1] – [3]. The priority needs to ensure greater safety for the patient and the need to facilitate health workers in the performance of specific activities can find effective answers in the computerization and automation of the process of drugs management.

It is widely acknowledged that the patient's safety and centrality in the care process are of interest to anyone dealing with health care and that the best approach to reducing errors in therapy is multidisciplinary. This approach deals with the HIS that involve a number of roles and functions - doctors, nurses, pharmacists and the patient himself – and a set of related processes.

The error cannot be completely cancelled but it can still be strongly reduced with the use of technological tools, such as HIS [1] [4]. The rest of this paper is organized as follows. Section II describes the related works. Section III introduces our proposal named FID-MED. Section IV describes the clinical risk reduction in FID-MED. Section V addresses the cost control in the proposed HIS. Section VI describes a brief comparison with competing system. The conclusions close the article.

II. Related Works

International studies [1] [4] show how the introduction of HISs contributes to general errors reduction even if they also warn against errors related to the use of system.

Developers of health care software have attributed improvements in patient care to these information systems. Like any health care intervention, such claims require confirmation in clinical trials.

The authors of [4] reviewed controlled trials assessing the effects of HISs by searching the MEDLINE, EMBASE, Cochrane Library, Inspec, and ISI databases and consulting reference lists through September 2004. Very briefly, the results are that the HISs improved practitioner performance in 62 (64%) of the 97 studies assessing this outcome, including 4 (40%) of 10 diagnostic systems, 16 (76%) of 21 reminder systems, 23 (62%) of 37 disease management systems, and 19 (66%) of 29 drug-dosing or prescribing systems. Fifty-two trials assessed 1 or more patient outcomes, of which 7 trials (13%) reported improvements. Improved practitioner performance was associated with HISs that automatically prompted users compared with requiring users to activate the system (success in 73% of trials vs 47%) and studies in which the authors also developed the HIS software compared with studies in which the authors were not the developers (74% success vs 28% respectively).

While several studies [1] - [3] deal primarily with injuries associated with errors in health care, the cost of inefficiencies related to errors that do not result in injury are also great. The effort associated with "missed dose" medication errors is just an example: when a medication dose is not available for a nurse to administer and a delay of at least two hours occurs or the dose is not given at all [6]. Nurses spend a lot of time tracking down such medications. Such costs are harder to assess than the costs of injuries and they may be even greater.

It is easy and common to blame operators for accidents but investigation often indicates that an operator "erred" because the system was poorly designed. In fact, machines can also produce errors or can mislead operators, e.g., if two medications that are spelled similarly are displayed next to each other, substitution errors can occur. Humans and machines are rather different and the combination of both has greater potential reliability than either alone. How best to make this synthesis is a very real problem. Humans are erratic, and err in surprising and unexpected ways; they are also resourceful and inventive and they can recover from errors in creative ways. In comparison, machines are stupid, more predictable but they cannot deal with a variable that was never anticipated, so that there was never any basis for equations to predict it or computers and software to control it.

As cited in [1], the system analysis [7] of a large series of serious medication errors (those that either might have or did cause harm) identified 16 major types of system failures associated with these errors. Of these system failures, all of the top eight could have been addressed by better medical information, e.g., laboratory systems do not communicate directly with pharmacy systems; prescribing, dispensing, and administering systems are not integrated, etc. Again, alerts are not delivered to caregivers in a timely way, while an increasing number of HISs contains data worthy of generating the alert message and many computerized physician order entry systems lack even basic screening capabilities to alert practitioners to unsafe orders relating to overly high doses, allergies, and drug–drug interactions.

An interesting side effect in HISs adoption is the awareness of the magnitude of problem human errors. In fact, [8] reports that while 92 % of hospital CEOs reported that they were knowledgeable about the frequency of medication errors in their facility, only 8 % said they had more than 20 per month, when in fact all probably had more than this.

Finally, several barriers have prevented implementation of these systems. Among these, the tendency of health care organizations to invest in administrative rather than clinical systems, the issue of "silo accounting" and the complexity of an effective systems integration due to the lack of standards and the difficulties with regard to the disclosure of data with third-party systems.

III. Our Proposal: FID-MED

The proposed HIS, named FID-MED, place itself in the sector of Health digitization and in particular of the computerized management of drugs, a growing sector seeking effective solutions. FID-MED is compliant with the main international medical standards, such as HL7 and DICOM, and can be easily integrated with third-party stocks management systems. It is oriented to the Electronic Health Record (EHR) and Electronic Medical Record (EMR) and it is compliant with Italian Personal Health Record (Fascicolo Sanitario Elettronico [9]).

FID-MED is mainly composed of a clinical Java web based application, named FIDCARE based on Oracle database, Figure 1, and an expressly designed medical cart, called FIDRUN, Figure 2. FIDCARE and FIDRUN work in synergy together and they are closely integrated. FIDRUN is equipped with PC on board with Monitor Touch 10.1" that allows accessing FIDCARE by Wi-Fi connection. FIDRUN has two series of drawers, one on the front (A side) and the other on the back (B side). Each series of drawers is divided into 8 configurable modules. Each module can be configured to with 6 small drawers or alternatively with 2 large drawer. FIDRUN represents a very valuable tool in the drug preparation and administration phases and has an operating autonomy that covers the working shifts.

A simplified deployment diagram of FID-MED is represented in Figure 3.

IV. Clinical Risk Reduction

FID-MED aims to reduce primarily the clinical risk through a tight control over the supported clinical processes (in accordance with the Italian applicable law) and consequently it works toward an efficient resources management and cost control. Mainly, FID-MED mitigates the clinical risk by means of:

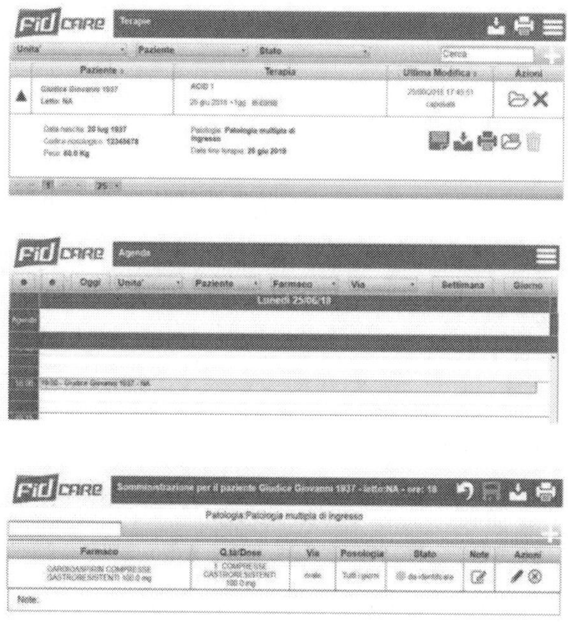

Figure 1. FIDCARE: Therapy prescription and administration pages.

- The tight control of the association of drugs to patient during the prescription, dispensing, and administration phases;
- The correct identification of the patients and the correct identification of the drugs in their therapies before drugs administration;
- The recognition and unique identification of the nurse who administers the therapy and the patient.

Once a physician prescribes a medicine to a patient through the computerized order entry, even if that medicine will be prepared (if needed) and administered by other operators (pharmacist or nurse) none of them can change drug, dose or the sequence without the physician permission. Through the correct identification of the patient and the drugs of his/her therapies, it is more difficult to be mistaken and the identification of the nurse who administers the therapy enforces the traceability of drugs from the physical order entry to the administration. After taking medications, the physician may perform a follow-up of your drug intake by displaying the patient's card.

The identification of patients, operators (physician, nurse, pharmacist, etc.) and drugs plays a central role in clinical risk mitigation with FID-MED. Essentially, in FID-MED the identification may occur in two different ways, other than the trivial selection in a list.

A. Reading barcodes.

The patient receives a bracelet with his/her own new unique barcode created by FIDCARE in the admission phase and the nurse uses a portable barcode scanner to identify the patient. The nurse may also type the barcode, if he/she does not has a barcode scanner or if it is temporarily unavailable or not working. In addition, operators have a barcode uniquely assigned to themselves, typically included in their badges, and they can use it for login in FIDCARE or FIDRUN, other than using user and password.

Figure 2. FIDRUN with fours modules. The first two at the top have each six small drawers and two modules at bottom have each two large drawers.

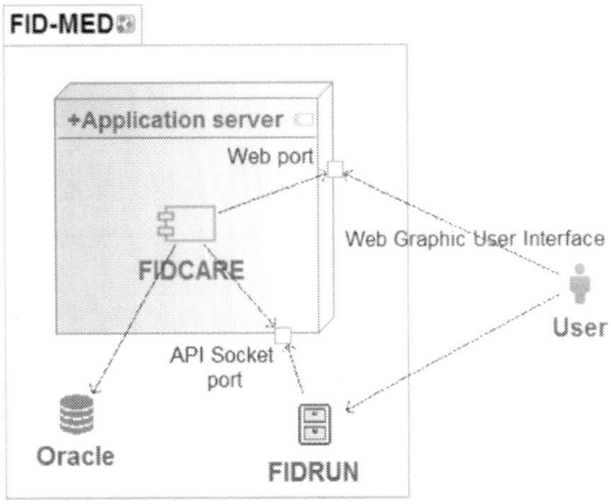

Figure 3. The deploy diagram of FID-MED.

All medicines have a barcode on the carton and it is quite easy to verify that, given a drug barcode, the corresponding drug belongs to a therapy (on that day, at that time) of a previously identified patient. A separate discussion is required for drugs that must be prepared before their administration, such as drips. In these cases, the package has not the barcode, and then FIDCARE creates a new unique barcode and print it on a sticker. The sticker includes details about the drugs, such as the chemical composition, the patient name and code, the preparation and expiration date and time, etc. From now on, FIDCARE will use this code to identify the drug, as in the previous case. Finally, barcodes can also be used to identify medical devices, such as stents. Here again, FIDCARE creates and prints a sticker.

B. Reading RFID tags

As in the barcode case, FIDCARE prints a new unique wearable passive tag during the admission and assigns it to the patient. Again, the nurse identifies the patient using a portable reader, and if necessary, the code may be manually typed. Substantially, RFID tags can be used in place of barcodes also for operators and medical devices identification.

These identification methods are reliable and fast, the main difference among them is that the reading of the RFID does not require the patient's vicinity to the nurse, unlike the barcode reading. It is worth mentioning that barcode and RFID technologies are not mutually exclusive. In fact, even if a drip may be tagged with RFID, it is not efficient to tag all drug's cartons due to the volume. A number of readers can read both barcode and RFID and FIDRUN is compliant with both of them. FIDRUN is equipped with this kind of reader connected via Bluetooth.

FIDRUN has two modes of operation: *patient mode* and *drug mode*.

In patient mode, before the administration of therapies, the web application requires an empty drawer for the patient and the cart automatically assigns and opens it. The nurse loads this *therapy drawer* with the drugs of the therapies of the patient and closes it. Then, when the patient is identified, the FIDRUN automatically opens the corresponding therapy drawer to allow the administration of the drugs. This operating mode contributes to the reduction of the risk of administration of a drug not present in the therapy of the patient. By default, the FIDRUN allows the opening of a single therapy drawer to further reduce the risk of administration of a wrong drug.

In drug mode, the FIDRUN automatically assigns a drawer, called the *drug drawer*, to each drug. The nurse pre-loads the cart by reading the barcode of a drug. If FIDRUN already assigned a drawer to the drug then it opens that drawer, otherwise FIDRUN assigns a new empty drawer. During the administration phase, the cart automatically opens the drug drawer if the corresponding medication is to be administered in a therapy. This case happens after the patient recognition and after the selection of a drug in its therapies in the web application. This mode of operation also reduces the risk of administration of the wrong drug.

Although these two modes of operation are described separately, FIDRUN can operate in mixed mode. In fact, it is possible to configure the medical cart to allocate a number of drawers to the patient mode and the remaining drawers to the drug mode. Finally, it is possible to use the drawers in a different way from the ones described above. This last operating mode, named *storage mode*, does not provide any form of drawer control by the cart and allows for greater freedom in their use.

V. Cost Control

As mentioned above, the control of cost is a matter of considerable importance in a HIS. FID-MED aims to controls the cost in multiple ways, among them:

- Introduction the concept of abstract drug;
- Optimization of chemotherapeutic medicinal products usage;
- Effective stocks management and usage product control.

A. *Abstract Drug*

Surprisingly, the introduction of the concept of *abstract drug* may affect significantly the cost related to drugs. An abstract drug has the following distinguishing features (pursuant to Italian legislation, DL n.95/2012):

a) *Active ingredient*: the active constituent of the drug. FIDCARE includes more than 1.100 active ingredients;
b) *Dosage form* (and its unit of measure) *and concentration* (and its unit of measure): e.g., *8% 100ML; 5 ml 400 UI/ml.*
c) *Pharmaceutical form*: the physical characteristics of the combination of active substance and excipients (non-active ingredients) forming a medicinal product (*tablet, liquid, capsule, gel, cream, sprays*, etc.). FIDCARE includes more than 150 pharmaceutical forms.

From a conceptual point of view, an abstract drug includes all information required for a physician order entry, i.e., a physician may insert a therapy using only abstract drugs. Of course, FIDCARE allows for commercial drugs selection if required. FIDCARE includes a list of about 2.500 abstract drugs and 3.200 commercial drugs. The association between abstract drugs and commercial drugs is one-to-many, that is, an abstract drug contains links to more than one commercial drug (having its active ingredient, dosage and pharmaceutical form) and a commercial drug is related to one abstract drug. Two commercial drugs are considered *equivalent* if they links the same abstract drugs. Under this assumption, before the administration, each abstract drug in a therapy can be substituted with one of the commercial drugs linked to it and FIDCARE allows a set of configurable rules with increasing degree of automation to do this. Given an abstract drug, the system can:

1) Require the explicit manual selection of the commercial drugs;
2) Choose the commercial drug having some (configurable) properties, for instance: bigger stocks, stocks as close as possible or any other rule. In this case, FIDCARE requires physician's confirmation to continue;
3) Same as previous one without confirmation.

The introduction of abstract drugs leads to a better drugs management because it allows a *separation of functions*. The physician can select an abstract drug without worrying about the policy on drugs stocks. Please, note that it is also possible to replace a commercial drug with an equivalent one. For instance, during the administration a nurse discovers that a drug package is empty. To complete the therapy, an equivalent commercial drug is needed and FIDCARE allows for the substitution after physician confirmation.

Finally, FIDCARE can automatically extract abstract drugs from a commercial drugs list. It analyses each commercial drug and extracts the abstract drug (if it does not exist) and creates the link between them. There are a lot of commercial drugs databases that FIDCARE can connect with, e.g., *Farmadati Italia s.r.l.* (Farmadati.it).

B. *Chemotherapeutic Medicinal Management*

The chemotherapeutic medicinals are among the most expensive medicinal products and their effective management can affect costs in a considerable way. From a point of view a little more formal, such medicinal can be seen as non-renewable resources that should be allocated to patients. Very often, such products may be used within a relatively short time-frame once the foil sachet is opened. The inefficiencies derive therefore from the non-use of products before their expiration. The optimization problem consists in the optimal planning of the therapy of the patients in order to minimize the non-use of the resources. The solution is a scheduling of therapy that also takes into account some constraints. Several constraints are: time constraints on therapies; patients preferences (days, times, etc.), the number of chairs available for patients, staff shifts, etc. This optimization problem falls into a family of problems very well studied in the scientific literature and FIDCARE implements a genetic algorithm specifically developed.

C. *Stocks management and usage product control*

Stocks management and therapies management are processes strictly correlated. As mentioned above, the abstract drug represents a separation element among them and the tight control over therapies scheduling provided by FIDCARE allows for a reduction of the stocks of the departments and of the warehouse. Furthermore, FIDCARE provides an Enterprise Data Warehouse module for reporting and data analysis. FIDCARE exposes a set of pre-set reports but users who are familiar with data Warehouse tools can easily create their own reports. Reports allow analysing the consumption of drugs by department, by period, by pathology, Figure 4. Based on these data, FIDCARE provides also a number of features for automatic reordering of stocks. The cost of therapies is inclusive of a number of cost items: drugs, equipment, staff, structure, etc. The report on the use of medications and medical devices also includes non-use due to: expiration date, breakage and damage, rejection by the patient, loss, etc. Finally, FIDCARE is designed to integrate with third-party stock management systems.

It is important to note that FIDCARE provides *actual* data on the consumption of drugs and devices. Otherwise, these data must be *estimated* using specific methodologies, such as Activity Based costing (ABC) or must be collected manually from paper documents. In the first case, it is inaccurate data while in the second case a substantial effort is required.

VI. COMPARISON WITH COMPETING SYSTEMS

This section describes a brief comparison with competing system having similar features. From an in-depth market analysis, the main competitors of FID-MED, on the Italian territory, are the computerized medical carts of the companies SPID S.p.A., IPSA s.r.l. and Max Medical Group s.r.l., three companies located in the North of Italy.

The cart of SPID S.p.A. allows the computerized management of the process of administering the drugs in the department. The system is the result of the combination of a computerized drugs trolley and a software for the prescription/administration of pharmacological therapies, integrated with an identification system.

The identification system, based on barcode or RFID technology, allows to identify the individual operator, the patient and the drug to be administered. Unlike FIDRUN, the cart of SPID unlocks the single drawer on operator's command. FIDRUN unlocks drawers automatically as the clinical processes require even if a drawer can be also unlocked on operator's request, if needed.

The *Smart Cart* of IPSA s.r.l, associated with a software for the management of therapies, allows the optimized management of all the phases of the pharmacotherapy process directly in patient bed and guarantees the traceability of the drugs and the activities of prescription and administration.

Smart Cart guides nurses in the distribution of therapies and allows automatic identification of operators, patients and medications by portable wireless barcode/RFID readers, finger-print readings, smart-cards. It is equipped with PC with Monitor Touch 10.1", medical keyboard, Wi-Fi, Bluetooth, long-life battery and a system of an access mechanism controlled to the contents of the drawers. Unlike FIDRUN, the Smart Cart does not allow multiple operating modes (patient, drugs and storage modes).

The automated cart of Max Medical Group s.r.l, named *Click*, uses a software for the management of prescription/administration. Click is equipped with an automatic mechanism of opening drawers and a control system that allows the nursing staff to have access to the only drawer containing the drug to be administered, while the other drawers remain locked. It is possible to configure the drawers according to the principle of assignment bed-drawer or alternatively active ingredient-drawer. FID-MED provides a more flexible assignment patient-drawer and a strong patient recognition mechanism.

FID-MED has some other interesting distinguishing features, among them: Voice User Interface, Gesture Interface, Adaptive User Interface, IoT Architecture and API.

A. *Voice User Interface*

Users interact with FIDCARE and FIDRUN through a voice in order to perform some tasks, e.g., initiate a new therapy, confirm an administration an automated, etc. Furthermore, FIDRUN gives acoustic feedbacks for a number of functionalities (e.g., correct execution, alerts, errors, etc.).

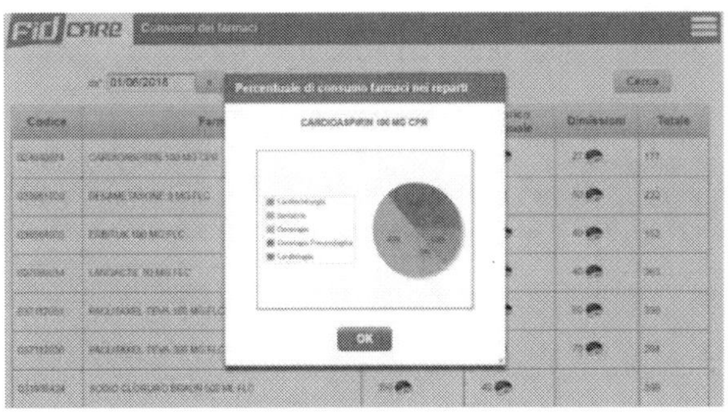

Figure 4. In the background, drugs consuption by period and by usage (therapies, manually unloaded, dismissal therapies). In the foreground, Cardioaspirin 100 mg consuption chart by departments.

B. Gesture Interface

Gesture recognition is achieved by FID-MED through small wearable three axis accelerometers for users. These sensors allow detection of a set of user gesture, such as move left/right and move top/down. The functionalities associated to that gesture (e.g., close, go to detail) can be customized.

C. Adaptive User Interface

FIDCARE adapts user interface in a light way. In particular, if the user must select a value from a list, it presents the preferred and most common previous choices of the user. For instance, this happens with drugs list, therapeutic protocols list, pathologies list.

D. IoT Architecture and API

FIDRUN implements a three levels IoT paradigm and, as previously described, it is deeply integrated into FID-MED but it is designed to be used in conjunction with third-party software as well. For this purpose, a set of Application Programming Interfaces (API) accessible via JSON messages on sockets or through Web Services is available. Through these APIs all capabilities of FIDRUN can be used in a programmatic way with any programming language (as FIDCARE do).

VII. Conclusion

This article presented FID-MED platform, a HIS composed by a Java web based application, named FIDCARE, and an especially designed integrated medical cart, called FIDRUN.

FID-MED is co-funded by Ministry of Economic Development START-Up 46/82 FIT DM 7/07/2009 (FID-MED 1.0) and Horizon 2020 – PON 2014/2020 and Campania Region Programma Operativo Regionale (POR) Fondo Europeo di Sviluppo Regionale (FESR) 2007/2013 (FID-MED 2.0). The article illustrates how the FID-MED platform may support the mitigation of the risk in healthcare activities leveraging on a tight integration between hardware and software technologies. Other than the complete control over all clinical processes (prescription, administering, stocks management, etc.), comparisons with competing system shows that FID-MED has some very interesting distinguishing features, such as voice and gesture user interface and easy integration with third party systems by design. FID-MED platform passed successfully several experimental usage trials to the hospital "Policlinico di Napoli". In this trial, FID-MED was integrated with the pre-existing acceptance module in the hospital from which it receives all information about hospitalized admissions. At present, the hospital in Pozzuoli uses FIDRUN in its urology department.

References

[1] L. T. Kohn, J. M. Corrigan, and M. S. Donaldson "To Err Is Human: Building a Safer Health System." Washington, DC: National Academy Press, 1999.

[2] C. J. McDonald, M. Weiner, and S. L. Hui "Deaths due to medical errors are exaggerated in Institute of Medicine report". JAMA. 2000;284: pp. 93–95.

[3] L. L. Leape, et al. "The nature of adverse events in hospitalized patients: results from the Harvard Medical Practice Study II". N Engl J Med. 1991;324(6): pp. 377–384.

[4] A. X. Garg, et al. "Effects of computerized clinical decision support systems on practitioner performance and patient outcomes a systematic review". JAMA.2005;293(10): pp. 1223–1238. doi:10.1001/ jama. 293.10.1223, March 2005.

[5] J. H. Gurwitz, et al. "Incidence and preventability of adverse drug events in nursing homes." Am J Med. 2000;109(2): pp. 87–94.

[6] D. W. Bates, D. L. Boyle, M. B. Vander Vliet, J. Schneider, and L. L. Leape "Relationship between medication errors and adverse drug events." J Gen Intern Med. 1995;10(4): pp. 199–205.

[7] L. L. Leape, et al. "Systems analysis of adverse drug events." ADE Prevention Study Group. JAMA. 1995;274(1): pp. 35–43.

[8] Bruskin-Goldring Research. (1999, February). B-D.id and CAP Today Co-Sponsor Survey Among Hospital Executives to Determine Awareness of Medication Administration and Specimen Collection Errors, from http://www.bd.com/contentmanager/b_article.asp?Item_ID=21202&ContentType_ID=1&BusinessCode=20001&d=&s=press&dTitle=Press&dc=Worldwide&dcTitle=Worldwide. Rerieved: August, 2018

[9] nsis.salute.gov.it. Access date: August, 2018

Depression Severity Prediction by Multi-model Fusion

Bin Li, Jie Zhu, Chunpeng Wang
Department of Electronic Engineering, Shanghai Jiao Tong University, Shanghai 200240, China
Emails: lb0621@sjtu.edu.cn, zhujie@sjtu.edu.cn, 2995078502@qq.com

***Abstract*—Depression is a common mental disorder and over 300 million people are estimated to suffer from it. Current main methods to predict the severity of depression depend on clinical interviews. Although very useful, this method lacks objectivity and efficiency. In this paper, we propose a multi-model fusion framework to detect the depression severity based on the random forest machine learning algorithms, and using features selected from different mediums. We first selected features of audio, video, and text contents of each patient with a fusion strategy using the decision-level fusion. The multi-model fusion regressors were trained with the extracted features to obtain the depression severity. To handle the imbalanced dataset problem, a sampling strategy was also conducted. Experiments were demonstrated on the Distress Analysis Interview Corpus-Wizard of Oz (DAIC-WOZ) database. The proposed framework achieved a Root Mean Square Error (RMSE) of 5.98, which is less than the baseline of 6.62 and suggest that our approach is efficient and suitable for the depression severity detection.**

Keywords–Depress detection; multi-model; fusion; machine learning.

I. Introduction

Depression is one of the most common psychological disorders, and a leading cause of disease burden [1]. According to the World Health Organization (WHO), over 300 million people are estimated to suffer from depression, i.e., $\sim$ 4.4% of the world's population [2]. As a global disease, the number of patients keeps increasing, especially in lower-income countries, where most of citizens are living to their age of high possibilities to have depression or other mental diseases. In addition to the persistent feelings of negativity, sadness, loss of interest or pleasure, and low self-esteem, depression is also corresponding to a range of physiological symptoms such as weight loss, insomnia and fatigue. It may be associated with diseases like bipolar disorder dementia [3] and even cardiovascular conditions [4]. Depression is never caused by only one thing, in the other words, a combination of factors such as biological factors, psychological factors and stressful life could be the reasons for depression. It has been identified as a burden to the economy while it affects justice and social systems [5]. Furthermore, the severe depression may lead to commit suicide and substance abuse [6]. Therefore, the detection and treatment of depression and its severity is of a high priority.

To measure the severity of depression, the Distress Analysis Interview Corpus (DAIC) is constructed, which contains clinical interviews in English for the diagnosis of psychological distress conditions such as anxiety, depression, and post-traumatic stress disorder [7]. The interviews were conducted by an animated virtual interviewer called Ellie (see Fig. 1), where she asked the interviewee a series of open-ended questions with intent of identifying clinical symptoms (see Fig. 2 for an example). The corpus includes audio and video recordings and extensive questionnaire responses. Each interview includes a depression score from the eight-item Patient Health Questionnaire depression scale (PHQ-8), which is established as a valid diagnostic and severity measure for depressive disorders in large clinical studies [8]. The PHQ-8 is a set of eight short multiple choices questions and every choice is corresponding to a score ranging from 0 to 3. Thus, the total point ranges from 0 to 24. A threshold of PHQ-8 score $\geq$ 10 is set to define current depression severity.

To predict the depression severity, many approaches have been proposed based on signal processing, computer vision, and machine learning algorithms. For instances, vocal utterances were taken advantaged to identify depression in [9]–[12], and intensive studies have been taken into facial expressions for the diagnostic of depression [13] [14]. The other researchers try to analyze depression from the text or language information [15] [16]. These approaches obtained results to some extent. The performance of using only one category feature is limited. They did not take full advantage of information we could get from the people we want to diagnose.

On machine learning based approaches, the decision-level fusion from multi-model strategy have been widely applied and proved efficient and accurate. Features were extracted from multiple mediums such as audio, video and text. Senoussaoui et al. proposed an i-vector based representation by extracting short term audio features for depression classification and prediction [17]. In [18], they used five acoustic feature categories, including prosodic, cepstral, spectral, glottal and, Teager Energy Operator (TEO) based features, to detect depression. Above approaches extracted features from one medium. In our view, features from multiple mediums may benefit the training of the regressors, which has been proved efficient in many works [19]. For the regressors for estimating the depression severity, some machine learning algorithms like support vector regressor (SVR), and random forest (RF) are applied. And the fusion or boosting by combining multiple models have been suggested with increase of those single model [20]. In [20], Morales and Levitan provided investigation of speech versus text features for depression detection systems, finding that multi-model system leads to the best performance. They also used automatic speech recognition to transcribe speech and found text features generated form ASR transcripts were useful for depression detection. In this paper, we propose to build a multi-model fusion framework, which fuses features extracted from audio, video, and text in the DAIC, and detect

Figure 1. Ellie, the virtual interviewer

```
Wizard-of-Oz
Ellie Who's someone that's been a positive
influence in your life?
Participant Uh my father.
Ellie Can you tell me about that?
Participant Yeah, he is a uh
Participant He's a very he's a man of few words
Participant And uh he's very calm
Participant Slow to anger
Participant And um very warm very loving man
Participant Responsible
Participant And uh he's a gentleman has a great
sense of style and he's a great cook.
Ellie Uh huh
Ellie What are you most proud of in your life?
```

Figure 2. Sample excerpts from virtual interviews

depression severity by predicting the PHQ-8 score of the patients.

This paper is organized as follows. In Section II we describe the selection of audio, video, and text features. The multiple-model fusion framework is explained in Section III. Experiments and results are demonstrated and discussed in Section IV. We conclude in Section V with outlooks.

II. Feature Extraction and Selection

The DAIC-WOZ dataset provides a set of samples to train the regressor, as well as features from audio, video recordings, as well as questionnaire responses in text format. However, there are two problems that should be considered before we begin our work. One is that the dataset is imbalanced, which has been found affecting the performance of machine learning algorithms [21]. In the DAIC-WOZ dataset, the number of samples is larger for the depressed than the non-depressed, which also leads an imbalanced distribution of the PHQ-8 scores. The distribution of DAIC-WOZ dataset in training and development sets is provided in TABLE I.

TABLE I. Distribution of DAIC-WOZ dataset in training and development sets

Dataset	*Depress*	*Non-depress*	*Sum*
Training	77	30	107
Development	22	13	35

TABLE II. Audio Features

Type	*Feature Name*
Prosodic	Fundamental frequency (F0), voicing (VUV)
Voice Quality	Normalised amplitude quotient (NAQ), quasi open quotient (QOQ), the difference in amplitude of the first two harmonics of the differentiated glottal source spectrum (H1H2), parabolic spectral parameter (PSP), maxima dispersion quotient (MDQ), spectral tilt/slope of wavelet responses (peak-slope), and shape parameter of the Liljencrants-Fantmodel of the glottal pulse dynamics (Rd)
Spectral	Mel cepstral coefficients (MCEP0-24), harmonic model and phase distortion mean (HMPDM0-24) and deviations (HMPDD0-12)
Formants	The first 5 formants

To handle this problem, we random oversample on the minor depressed dataset, to make the numbers of samples of every PHQ-8 score be roughly the same. The other problem is that the sample size is small, the number of samples in training set is only 107. On account of such a small sample size, the number of features should also be small to avoid dimensionality and overfitting. So, the features to be used should be filtered and selected, since many of them are with very small values approaching to zeros.

A. Audio Features

Many researches have shown that the speech production of a human is very complex and as a result slight cognitive or physiological changes can produce acoustic changes [22]. Depressed speech is always associated with a wide range of prosodic, source, formant and spectral indicators. The audios provided by DAIC-WOZ dataset consist of a series of features extracted by the Covarep toolbox [23] at 10 ms intervals over the entire recorded wav files. This toolbox can capture both prosodic characteristics of the speaker, as well as voice quality. All audio features available for this dataset are listed in the Table II. One thing should be noticed is that the VUV(voiced/unvoiced) provides a flag (0, 1) to represent whether the current segment is voiced or unvoiced. If it is unvoiced, the other features are not utilized. In order to uncorrelated with speaker, the F0 was normalized to range from 0 to 1, and the deltas and delta-deltas were extracted for F0 and MCEPs.

We use the mean and standard deviation of the raw voice features during voice time. So, there are totally 252 different kinds of audio features for us to choose from. The raw recorded wav files of the virtual interview are provided, so we can extract other features for further analysis.

B. Video Features

Video features have been widely used for depression analysis, including facial subtle expression, body movement, gestures and facial muscle movements. Facial expressions can be an extremely powerful medium used to convey human overt emotional feedback. Girard et al. found that people with high level depression made fewer affiliative facial expression and more non-affiliative facial expression [24]. Facial Action Coding System (FACS) is a system to taxonomize human facial movements by their appearance on the face. Movements of individual facial muscles are encoded by FACS from slight different instant changes in facial appearance. By using FACS, it is possible to code nearly any anatomically possible facial expression, deconstructing it into the specific Action Units (AU) that produced the expression. It is a common standard to objectively describe facial expressions. The DAIC-WOZ database provides different types of video features based on the OpenFace framework [25], such as facial landmarks, HOG (Histogram of Oriented gradients) features, gaze direction estimate for both eyes, head pose and AUs (Action Units). In this paper, we use the mean and standard deviation of 20 key AUs over time. Thus, we get 40 video features for use.

C. Text Features

Clinical psychologist diagnose depression is mainly base on the language. We use speech features because the cognitive and physical changes associated with depression can lead to differences in speech. Linguistic features are similar to this, psychological and sociological theories suggest that depressed language can be characterized by specific linguistic features. The DAIC-WOZ database provides us the transcript file of the interview. Since the provided transcripts are human-machine conversations, we first extract human part. Then, we use Linguistic Inquiry and Word Count (LIWC) [26] software to count the frequency of word for each interview. The LIWC software is developed for providing an efficient and effective method for studying the various emotional, cognitive, and structural components present in individuals' verbal and written speech samples. With the help of LIWC, we extracted 93 text features. In order to get more text features of value, we use the depression related words list available online and the AFINN [27] toolbox. The depression related word list contains more than 200 words, such as "Abnormal", "Alone"and "Anger". The AFINN toolbox would represent the valence of current text by comparing it to given word list with known sentiment labels. The output scores of AFINN is between minus five (negative) and plus five (positive). The mean, median, minimum, maximum and standard deviation of the output scores were used. So, we get five additional text features.

D. Feature Selection

Now, we already get the feature we want to use. We also should take the gender into consideration. In [28], Stratou et al. showed that gender plays an important role in the assessment of psychological conditions such as depression and PTSD, and a gender dependent approach significantly improves the performance over a gender agnostic one. The TABLE III shows the dimension of each feature category in the feature vector.

TABLE III. Dimension of each feature category

Feature Name	*Dimension*
COVAREP	252
Formants	10
AUs	40
LIWC	93
AFINN	5
Gender	1
Sum	401

We totally get 401 features to use. However, there are only one score need to be predicted for each example. What' s more, the number of the train set samples is only 107, among them the number of depression samples is 30. And only a small number of features are actually useful. So, it' s essential to small the number of features to avoid the problems of dimensionality, overfitting and shorter train times. First, we discard some features that always zero, such as HMPDM_0-3, and features with too much abnormal values, such as HMPDM_4. Next, we remove the feature with low variance according to the statistics of every feature. Then, we conduct the L1-based feature selection method. Linear models penslized with the L1 norm have sparse solutions, many of their coefficients are zero.When the goal is to reduce the dimensionality of the data, they can be used to select the non-zero coefficients. Finally, we select the best feature set.

III. Multi-Model Fusion

We use the PHQ-8 score as description of the depression severity. The distribution of the depression scores based on the train and development set is shown in Figure 3 [29]. As we can see from the figure, the data are imbalanced, which mean there are much more samples with low PHQ-8 scores than those with high PHQ-8 scores. It has been reported that imbalanced classes of data will have a great influence on the performance of machine algorithms. So, we conduct a random oversampling to make the number of samples of each PHQ-8 score is roughly the same. Because the available data set are too small, a 5-fold cross-validation has been used. That mean we divided development set into 5 folds. And we use one fold for testing and another 4 folds for training each time. We conduct Root Mean Square Error (RMSE):

$$RMSE = \sqrt{\frac{\sum_{i=1}^{n}(y_i - \hat{y}_i)^2}{n}} \quad (1)$$

to evaluate the performance of the multi-model. While the RMSE score is lower, the performance of the multi-model is better.

Random Forest (RF) is an ensemble learning method which fits multiple decision tree regression by random

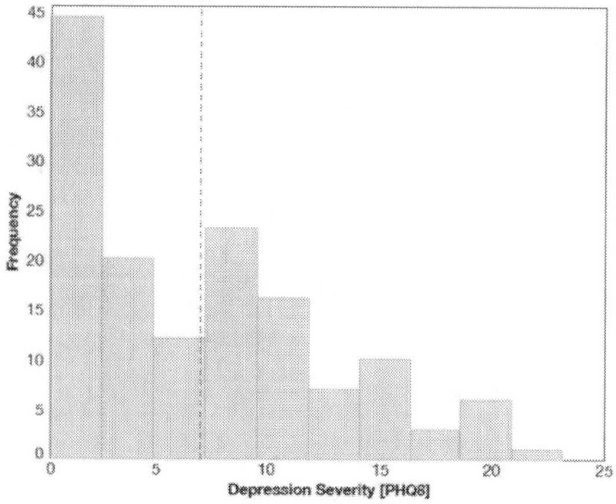

Figure 3. "Histogram of depression severity scores of training and development set"

selection of features and optimizes by bagging and aggregating the results. The random forest model is a type of additive model that makes predictions by combining decisions from a sequence of base models. Each base model is a sample decision tree. The main principle behind ensemble learning method is that a group of "weak learners" can come together to form a "strong learner". The random forest is of great advantages, such as fast training, high generalization capability and stable performance. Many similar researches related with depression severity prediction use random forest model. Support vector regression (SVR) uses the same principles as the support vector classification, with only a few differences. It just looks for and optimize the generalization bounds the same way as classification approach. In our work, we have tried a grid search for random forest regressor and support vector regression (linear kernel) as base regressor. However, the performance of random forest is better than support vector regression.

The framework of our proposed method is shown in Figure 4. As we mentioned above, gender is of great concern for depression severity detection. The performance of with or without gender will show in the next section. We firstly conduct a feature-level fusion, combining audio, video or text feature with gender. Then, three random forest regressors were constructed to calculate the regression value only depending on fused features. Finally, we implemented a decision-level fusion method. The linear opinion pool method was employed in the final decision-level multi-model fusion because of its simplicity. The fused result can be formulated as follow:

$$R_{final}(x) = \sum_{i=1}^{n} \alpha_i R_i(x) \qquad (2)$$

where x stands for a test sample, R(x) stand for the result of ith random forest regressor, α_i is the corresponding weight which satisfies.

$$\sum_{i=1}^{n} \alpha_i = 1 \qquad (3)$$

We use the optimal weights to fuse the regression value together to vote for a result that minimize error. The fusion result is shown in TABLE IV.

IV. Results and Discussion

We first needed to decide which base model to use. We have tried the random forest regressor and support vector regressor (linear kernel). The performance of this two regressor is provided in TABLE IV. In this experiment, we conducted very little optimization. Obviously, the performance of random forest is better than SVR in every feature. So, we selected the random forest regressor as base model. In the research of the other scientist found that fusing gender feature could achieve better performance. Therefore, we conducted a feature-level fusion, combining audio, video or text feature with gender. The performance of audio, video and text feature only and the performance with gender are provide in TABLE V. The weight of every base model is another problem we have to deal with. We first tried equal weight of every base model strategy. However, this strategy seemed that it could not reach the optimal result. So, we changed the weight of every base model according to their performance. Finally, we chose the weight 0.4 for audio, 0.4 for video, 0.2 for text.

TABLE IV. Performance of random forest and SVR

Features	*Base Model*	*RMSE*
Audio	Random forest	7.16
Video	Random forest	7.48
Text	Random forest	7.13
Audio	SVR	8.16
Video	SVR	7.81
Text	SVR	7.35

TABLE V. Performance include gender or not

Features	*RMSE*
Audio only	6.53
Video only	6.92
Text only	6.70
Audio&Gender	6.39
Video&Gender	6.89
Text&Gender	6.58

The AVEC 2017 organizer provided the baseline of Depression Severity Assessment Challenge (DSC). They computed the depression severity baseline using random forest regressor (number of trees: 10, 20, 50, 100, 200). In their experiment, the best performing random forest has trees = 10. Fusion of audio and video modalities was performed by averaging the regression outputs of the unimodal random forest regressors. The performance of RMSE is provided in TABLE VI.

We take the optimizing strategy mentioned above, the result of proposed multi-model fusion framework is displayed in TABLE VII. All these results are on the

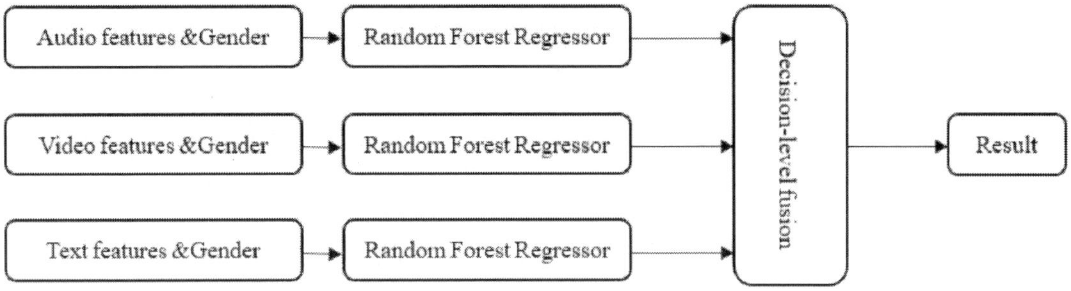

Figure 4. Framework of multi-model fusion

TABLE VI. Baseline of DSC challenge

Modality	*RMSE*
Audio	6.74
Video	7.13
Audio-Video	6.62

developing set. The result obtained were better than the baseline result.

TABLE VII. Performance of multi-model fusion framework

Features	*RMSE*
Audio&Gender	5.84
Video&Gender	5.89
Text&Gender	6.43
Multi-model fusion	5.98

V. Conclusion

Depression is a widespread mental disorder now. The accurate detection of it is still hard by manual. In this paper, we proposed a random forest regressor based multi-model fusion framework on audio, video and text features for prediction of PHQ-8 scores which ranges from 0 to 25. The weights of base models depend on their own performance. Our experiments suggested that the proposed approach performs better than DSC baseline. Since the result of multi-model fusion is very promising, future work would be devoted toward using more complex base regressor, such as neural networks. With more relevant features containing more useful information, the overall performance could also be improved. The features of body and head movement could be the next experiment we would like to try. And the more effective fusion strategy is another future work direction. The linear opinion pool method seemed could not reach the best performance of every base regressor. We may use some method to grade the performance of base model in every sample. Then, adjusting the weights dynamically to reach the best potential.

Acknowledgment

This work was supported by the National Key Research and Discovery Plan No. 2017YFF0210903, and the National Natural Science Foundation of China (grant No. 11433002).

References

[1] C. D. Mathers and D. Loncar, "Projections of global mortality and burden of disease from 2002 to 2030," PLoS medicine, vol. 3, no. 11, 2006, p. e442.

[2] WHO et al., "Depression and other common mental disorders: global health estimates," World Health Organization, 2017.

[3] V. Kral, "The relationship between senile dementia (alzheimer type) and depression," The Canadian Journal of Psychiatry, vol. 28, no. 4, 1983, pp. 304–306.

[4] R. M. Carney, K. E. Freedland, G. E. Miller, and A. S. Jaffe, "Depression as a risk factor for cardiac mortality and morbidity: a review of potential mechanisms," Journal of psychosomatic research, vol. 53, no. 4, 2002, pp. 897–902.

[5] P. E. Greenberg, A.-A. Fournier, T. Sisitsky, C. T. Pike, and R. C. Kessler, "The economic burden of adults with major depressive disorder in the united states (2005 and 2010)," The Journal of clinical psychiatry, vol. 76, no. 2, 2015, pp. 155–162.

[6] K. Hawton, C. C. i Comabella, C. Haw, and K. Saunders, "Risk factors for suicide in individuals with depression: a systematic review," Journal of affective disorders, vol. 147, no. 1, 2013, pp. 17–28.

[7] J. Gratch et al., "The distress analysis interview corpus of human and computer interviews." in LREC. Citeseer, 2014, pp. 3123–3128.

[8] K. Kroenke, T. W. Strine, R. L. Spitzer, J. B. W. Williams, J. T. Berry, and A. H. Mokdad, "The phq-8 as a measure of current depression in the general population." Journal of Affective Disorders, vol. 114, no. 1, 2009, pp. 163–173.

[9] N. Cummins, J. Epps, M. Breakspear, and R. Goecke, "An investigation of depressed speech detection: Features and normalization," in Twelfth Annual Conference of the International Speech Communication Association, 2011, pp. 2997–3000.

[10] S. Scherer, G. Stratou, J. Gratch, and L.-P. Morency, "Investigating voice quality as a speaker-independent indicator of depression and ptsd." in Interspeech, 2013, pp. 847–851.

[11] Y. Yang, C. Fairbairn, and J. F. Cohn, "Detecting depression severity from vocal prosody," IEEE Transactions on Affective Computing, vol. 4, no. 2, 2013, pp. 142–150.

[12] N. Cummins, V. Sethu, J. Epps, S. Schnieder, and J. Krajewski, "Analysis of acoustic space variability in speech affected by depression," Speech Communication, vol. 75, 2015, pp. 27–49.

[13] J. F. Cohn et al., "Detecting depression from facial actions and vocal prosody," in Affective Computing and Intelligent Interaction and Workshops, 2009. ACII 2009. 3rd International Conference on. IEEE, 2009, pp. 1–7.

[14] G. Stratou, S. Scherer, J. Gratch, and L.-P. Morency, "Automatic nonverbal behavior indicators of depression and ptsd: Exploring gender differences," in Affective Computing and Intelligent Interaction (ACII), 2013 Humaine Association Conference on. IEEE, 2013, pp. 147–152.

[15] M. De Choudhury, M. Gamon, S. Counts, and E. Horvitz,

"Predicting depression via social media." ICWSM, vol. 13, 2013, pp. 1–10.

[16] M. Valstar et al., "Avec 2016: Depression, mood, and emotion recognition workshop and challenge," in Proceedings of the 6th International Workshop on Audio/Visual Emotion Challenge. ACM, 2016, pp. 3–10.

[17] M. Senoussaoui, M. Sarria-Paja, J. F. Santos, and T. H. Falk, "Model fusion for multimodal depression classification and level detection," in Proceedings of the 4th International Workshop on Audio/Visual Emotion Challenge. ACM, 2014, pp. 57–63.

[18] L. Low, "Detection of clinical depression in adolescents' using acoustic speech analysis," researchbank.rmit.edu.au, 2011.

[19] S. Dham, A. Sharma, and A. Dhall, "Depression scale recognition from audio, visual and text analysis," arXiv preprint arXiv:1709.05865, 2017.

[20] M. R. Morales and R. Levitan, "Mitigating confounding factors in depression detection using an unsupervised clustering approach," in CHI 2016 Computing and Mental Health Workshop, 2016.

[21] X.-Y. Liu, J. Wu, and Z.-H. Zhou, "Exploratory undersampling for class-imbalance learning," IEEE Transactions on Systems, Man, and Cybernetics, Part B (Cybernetics), vol. 39, no. 2, 2009, pp. 539–550.

[22] N. Cummins, J. Joshi, A. Dhall, V. Sethu, R. Goecke, and J. Epps, "Diagnosis of depression by behavioural signals: a multimodal approach," in Proceedings of the 3rd ACM international workshop on Audio/visual emotion challenge. ACM, 2013, pp. 11–20.

[23] G. Degottex, J. Kane, T. Drugman, T. Raitio, and S. Scherer, "Covarep ¡ª a collaborative voice analysis repository for speech technologies," in IEEE International Conference on Acoustics, Speech and Signal Processing, 2014, pp. 960–964.

[24] J. M. Girard, J. F. Cohn, M. H. Mahoor, S. M. Mavadati, Z. Hammal, and D. P. Rosenwald, "Nonverbal social withdrawal in depression: Evidence from manual and automatic analyses," Image and vision computing, vol. 32, no. 10, 2014, pp. 641–647.

[25] T. Baltrušaitis, P. Robinson, and L.-P. Morency, "Openface: an open source facial behavior analysis toolkit," in Applications of Computer Vision (WACV), 2016 IEEE Winter Conference on. IEEE, 2016, pp. 1–10.

[26] J. W. Pennebaker, R. L. Boyd, K. Jordan, and K. Blackburn, "The development and psychometric properties of liwc2015," Tech. Rep., 2015.

[27] F. Å. Nielsen, "A new anew: Evaluation of a word list for sentiment analysis in microblogs," arXiv preprint arXiv:1103.2903, 2011.

[28] G. Stratou, S. Scherer, J. Gratch, and L.-P. Morency, "Automatic nonverbal behavior indicators of depression and ptsd: the effect of gender," Journal on Multimodal User Interfaces, vol. 9, no. 1, 2014, pp. 17–29.

[29] F. Ringeval et al., "Avec 2017: Real-life depression, and affect recognition workshop and challenge," in Proceedings of the 7th Annual Workshop on Audio/Visual Emotion Challenge. ACM, 2017, pp. 3–9.

An EHR Platform for Realizing Precision Medicine in Emergency Care

Vassiliki Koufi Flora Malamateniou George Vassilacopoulos

Department of Digital Systems
University of Piraeus
Piraeus, Greece
email: {vassok, flora, gvass}@unipi.gr

Abstract— **Precision medicine has emerged as a new field that facilitates the discovery of individualized diagnostic and treatment plans for patients based on the way their unique characteristics may dictate disease response or progression. Despite its potential, there are several settings where precision medicine remains more aspiration than reality. One such setting is emergency care where physicians are called upon initiating investigations and interventions to diagnose and/or treat patients in the acute phase and making decisions regarding a patient's need for hospital admission, observation, or discharge. In this effort, emergency physicians usually have little or no information at hand. The convergence of healthcare data with multi-omics and real-world data heralds a new era where an individualized approach to emergency care seems feasible. Especially, with the latest technological innovations, physicians can be provided with patient healthcare data and tools for the analysis of these data against the various genomically and phenotypically defined patients/citizens. This paper presents a cloud-based emergency care delivery platform that incorporates an Electronic Health Record (EHR) with the aforementioned features to meet the needs of emergency care practitioners. The platform aims at bridging the translational gap between bench and bedside and moving towards a realization of personalized and stratified medicine by discovering associations between disease and genetic, environmental or process measures.**

Keywords-EHR; precision medicine; emergency care; healthcare analytics.

I. INTRODUCTION

The explosion of precision medicine and genomics research over the last decade or more is expected to herald a rapid acceleration in the identification and development of next-generation prevention and therapeutic services [1][2]. Essentially, precision medicine constitutes a holistic approach to disease treatment and prevention that considers variability in people's genes, environment, and lifestyle [3][4][5]. Under this paradigm, interventions can be tailored to individuals or groups, rather than using a one-size-fits-all approach in which everyone receives the same care. To this end, several tools need to be employed such as molecular diagnostics, imaging, and analytics [2][3]. However, there are certain challenges to clinical implementation of precision medicine which have been identified at multiple levels. This paper is concerned with the challenges related to the technologies that need to be employed in order for precision medicine goals to be met in emergency care environments.

Nowadays, research and development directors in the life sciences are working with enormous, disparate data sets, which must be stored and analyzed accurately and quickly. These data may include genomic or proteomic data (large-scale studies of organic proteins), socioeconomic data, patient medical history data, scientific texts, clinical studies and data from universities and laboratories, all in different formats and from various sources [1]. Thus, intelligent insights from that diverse, big amount of data is really hard to obtain. An Electronic Health Record (EHR) that will integrate all the above types of data and will be enhanced with decision support tools can lay the foundation for the realization of precision medicine by providing clinicians and patients with individualized information and preferences, intelligently filtered at the point of care. This is particularly useful in the emergency care where a patient's life may hinge on the instant availability and accuracy of these pieces of information. In particular, in the case of patients already genotyped in an outpatient clinic, data stored in such an EHR system could be available when they present acutely ill to the Emergency Department (ED), while in the case of patients not previously genotyped, the tools provided by such an EHR system could assist in obtaining real-time results (i.e., in fewer than several hours) when these patients present to the ED. Thus, genome-guided decision making can be supported when it is mostly needed.

This paper presents a cloud-based emergency care delivery platform that incorporates a next-generation EHR, which can serve as a supporting tool to the implementation of both population- and individual-level interventions in emergency care settings. Thus, health benefits can be maximized, harm can be minimized, and unnecessary healthcare costs can be avoided. The EHR proposed incorporates a broader data set, compared to data sets in legacy or existing EHRs, including: (a) patient information contained in Personal Health Records (PHRs), (b) health information from medical devices connected to patient such as from assistive telecare systems, (c) social care information retrieved on request from social care organizations, (d) health information extracted from various healthcare systems such as primary and hospital care Electronic Medical Records (EMRs), (e) data regarding a person's lifestyle obtained from social media like Facebook; and (f) genomics information such as genotype and sequence data extracted from biobanks and genetic databanks. The privacy concerns arisen in such an infrastructure due to its multi-owner and multi-user nature are met by a Health Insurance Portability and Accountability Act (HIPAA) compliant access control mechanism, whereby fine-grained access control is provided to users [6]. This mechanism draws from the Attribute-based Access Control (ABAC) paradigm and supports HIPAA "break the glass" access to data pulled together from all the data sources. Thus, access to the EHR of a patient is automatically granted to all physicians and nurses on duty in the ED upon patient arrival and revoked upon patient departure from the ED unless stated otherwise.

The remainder of the paper is structured as follows: Section 2 presents related work; Section 3 describes the scenario that motivated our research; Section 4 presents the architecture of the proposed EHR platform; Section 5 describes the main issues related to the implementation of the proposed platform and Section 6 concludes the paper.

II. RELATED WORK

The last few years have seen huge leaps in several research fields, such as molecular biology, genomics, and bioinformatics, which, in turn, led, among the rest, to the emergence and continued growth of precision medicine [7]. This paradigm shift in care delivery has been further expedited by converging trends of increased connectivity, through social media and mobile devices, and citizens' growing desire to be active partners in their health and wellness management [7].

A key enabling factor for the realization of precision medicine in emergency care is the Information Technology (IT) infrastructure that will support it. The infrastructure envisaged here is a cloud-based emergency care delivery platform that incorporates an EHR which is enhanced with advanced analytics tools to provide the information required for producing useful insights regarding emergency care delivery. Hence, the infrastructure is meant to be incorporated into an Emergency Medical Service (EMS). The last few years, a number of approaches have been proposed according to which health records, either EHRs or PHRs, are stored on a Cloud and can be accessed via mobile devices. In particular, in [8], an integrated EMS cloud-based architecture is proposed that allows authorized users to access emergency case information in standardized document form, as proposed by the Integrating the Healthcare Enterprise (IHE) profile, uses the Organization for the Advancement of Structured Information Standards (OASIS) standard Emergency Data Exchange Language (EDXL) Hospital Availability Exchange (HAVE) for exchanging operational data with hospitals and incorporates an intelligent module that supports triaging and selecting the most appropriate ambulances and hospitals for each case. Moreover, Watson for Genomics is a cognitive system that integrates massive amounts of new omic data with the current body of knowledge to assist physicians in analyzing and acting on patient's genomic profiles [9]. In addition, N-of-One platform assists in linking each patient's unique profile to potential treatment strategies in context, supporting targeted clinical decisions [10]. Thus, clinicians are provided with concise, accurate and clinically meaningful interpretation of molecular test results [10]. The 2bPrecise platform captures and stores genomic data from a range of sources, harmonizes clinical knowledge and genomic research to identify relevant information, and then pushes the resulting insights to the point of care to enable action [11]. Finally, several approaches to EHR integration with PHRs and other information systems have been proposed throughout the last few years [12][13][14]. Compared to these approaches, the EHR-based platform proposed in this paper facilitates the integration of a wider range of information sources into the clinical workflow, thus enabling the derivation of deeper insights into several diseases. Moreover, the tools incorporated in it enables the analysis of the EHR data against the various genomically and phenotypically defined patients/citizens thus providing the ability for individualized emergency care.

III. MOTIVATING SCENARIO

The basic motivation for this research stems from our involvement in a recent project concerned with designing and implementing an EHR-based platform for the provision of medical data access at the point of care while fully protecting privacy. This involves providing access to the appropriate people, based on patient authorizations, but also granting access to the patient's data in cases where his/her involvement in the authorization propagation process is not feasible. Moreover, it involves the analysis of these data in order to provide insights that will assist emergency physicians in diagnosing and/or treating patients in the acute

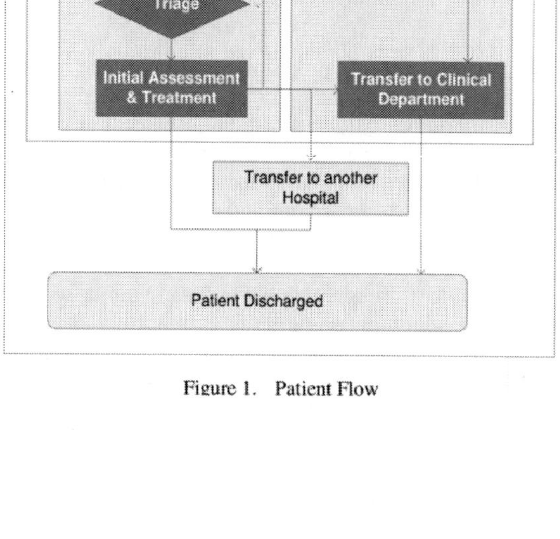

Figure 1. Patient Flow

phase.

Suppose a healthcare delivery situation that takes place within a health district where an individual is transferred to a hospital's ED (Figure 1). Upon arrival to the ED, the individual is registered as an emergency patient and undergoes a brief triage in order for the nature and severity of his/her illness to be determined. If his/her illness or injury is considered to be serious he/she is seen by a physician more rapidly than the patients with less severe symptoms or injuries. After initial assessment and treatment, the patient is either admitted to the hospital (e.g., to a clinical department or the Intensive Care Unit - ICU), stabilized and transferred to another hospital for various reasons, or discharged.

As many ED visits are unplanned and urgent, there is a need to ensure that information regarding the longitudinal patient health condition (e.g., problems, allergies, medications, diagnoses, recent procedures, recent laboratory tests) is conveyed to ED physicians automatically upon registration of a patient to an ED. Thus, inefficiencies in care, in the form of redundant testing, care delays, and less-effective treatments prescribed are eliminated and quality of care is enhanced.

IV. System Architecture

Figure 2 illustrates a high-level architectural view of the proposed EHR-based platform. As illustrated, the proposed platform comprises three layers, i.e., the resources layer, the services layer and the presentation layer.

The resources layer hosts the remote data resources providing the various chunks of data comprising a next-generation EHR. These resources are heterogeneous and reside at different settings. The various parts of a next-generation EHR are accessible by relevant web services, which hide away specific implementation details of each resource and provide a uniform and consistent interface to the operations in that resource. However, these resources are owned by different entities according to their specific security policy (e.g., owner of PHR and genomic data is the patient, while owner of the clinical/social data is the health/social provider where the patient has received care). Hence, each part of the EHR may be governed by different institutional or personal policies and practices with respect to confidentiality, security and release of information. However, as health and social care organizations constitute covered entities under the HIPAA Privacy Rule, they must comply with the Rule's requirements for safeguarding the privacy of protected health information.

The information stored in the proposed next-generation EHR are segregated into three categories, i.e., medical, lifestyle and genomic information.

Medical information may be obtained from:

- **EHRs:** They provide health information collected from healthcare providers where patients have received care in the past. As such, it is usually extracted from various healthcare systems such as primary and hospital care EMRs.
- **PHRs:** They provide patient health information contained in PHRs. This is mostly information from

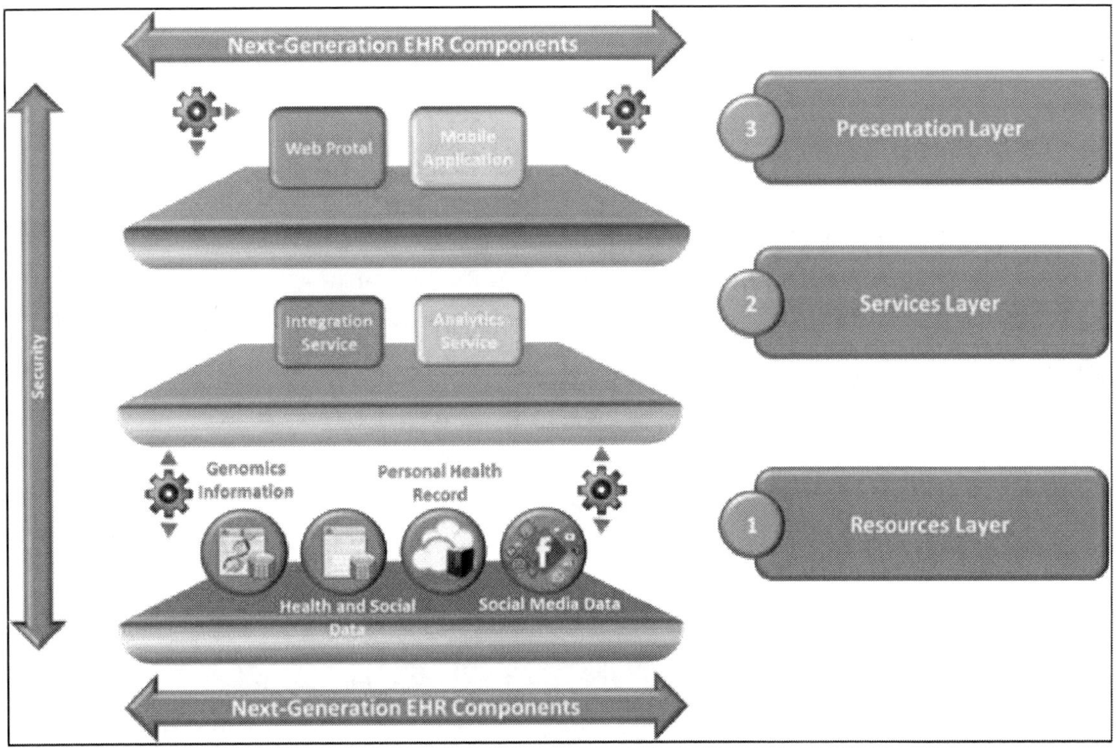

Figure 2. System Architecture

medical devices attached to patients and it is uploaded to the PHR through a special-purpose mobile application [15].

- **Social Care Records:** They provide information collected by social care organizations which have provided their services to a patient. It is retrieved on request from social care organizations.

For the purposes of this paper, it is assumed that PHR information is stored in a cloud-based PHR while EHR and social information is stored in data repositories of geographically distributed and organizationally disparate health and social care providers, respectively. Data regarding the patient's condition are either collected by the ambulance staff or provided by the patient himself upon arrival at the ED of the hospital. These data are stored at the ED information system.

Lifestyle information is obtained from **(a)** patients' PHRs, which, apart from the health information mentioned above, may include other types of information as well (e.g., exercise and dietary habits), and **(b)** patients' activity in social media (places they have visited, sleeping hours depending on the times they become active, etc.).

Genomic information is assumed to be hosted in several research centers. It may contain genotype and sequence data extracted from biobanks and genetic databanks. As molecular phenotyping is a time-consuming procedure, it would be particularly useful if a patient has already had a complete genomic or other systematic molecular analysis performed prior to his visit at the ED of a hospital. Currently, there is a rising number of validated clinical applications for molecular phenotyping; this leads to more patients having such analyses available from their usual clinical care [3].

The services layer, as indicated by its name, hosts a number of services pertaining to information integration and analysis. This layer resides in a trusted cloud and it, currently, hosts the following two cloud-based services. Moreover, it is scalable in the sense that it can be further enhanced with additional tools as long as they are implemented in the form of cloud-based services and published in this layer so that they become instantly available.

- The **Integration Service** serves as a mediation gateway that handles interactions between users and data resource providers. Typically, it provides access to integrated patient EHR data while ensuring compliance to the relevant HIPAA and patient-defined policies by properly authorizing users of the system, i.e., medical doctors. In doing so, the Integration Service draws upon the ABAC paradigm, i.e., it mediates between requestors (e.g., healthcare professionals) and resources (EHR web services) to decide whether access of a given requestor to given resource should be permitted or denied by taking into account the attributes of the requestor, the resource, the action and the context holding at the time of the attempted access (operational, technical, or situational). However, in the case of an emergency, ED doctors gain access to a full set of a patient's information through a "break the glass" policy supported by the system.
- The **Analytics Service** constitutes a recommender system that utilizes the collaborative filtering technique in order to associate each patient arriving at the ED of a hospital with other patients who have sought emergency care in the past due to experiencing similar conditions.

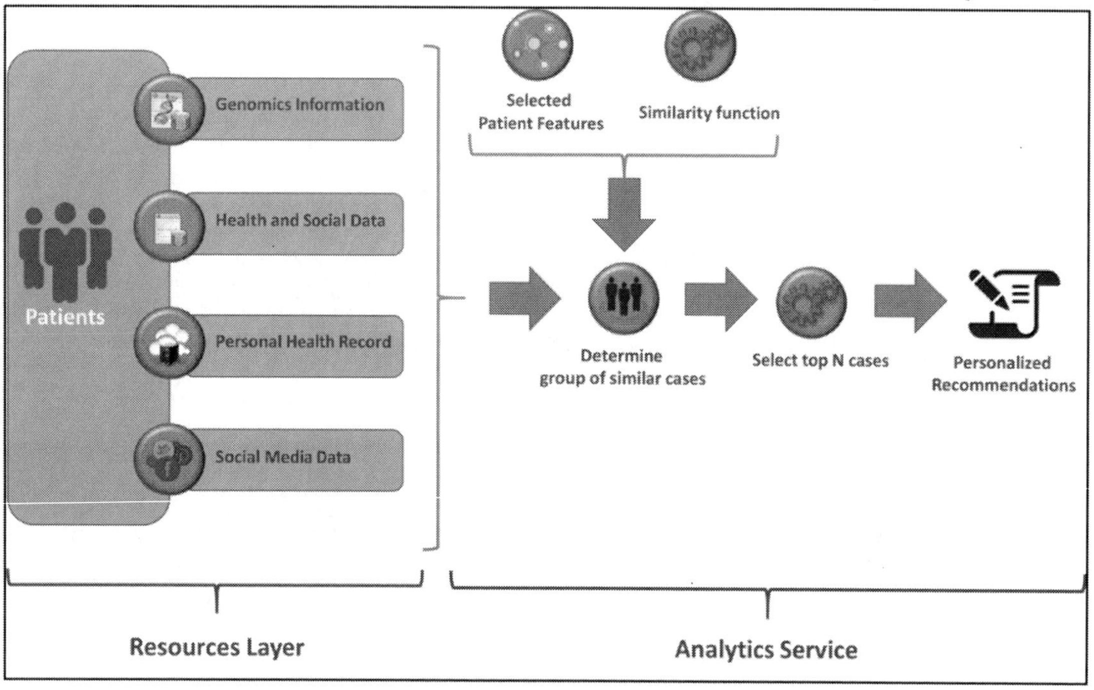

Figure 3. Analytics Service

To this end, certain features of patients are taken into consideration, which are retrieved from the next-generation EHRs (e.g., shared diseases, symptoms, family histories, lab results, urban/rural residencies, occupation, demographics, etc.). Thus, a subset of patients is identified who appear to resemble the patient currently under treatment in the ED of a hospital. Based on these cases, personalized recommendations can be extracted regarding the optimal treatment of the current patient. These recommendations are based on the outcome of treatments which have been used in similar cases in the past and are provided to ED physicians. Figure 3 illustrates a high level view of the Analytics Service incorporated in the proposed system.

The presentation layer provides the user interface whereby emergency physicians can gain full access to information included in next-generation patient EHRs. To this end, both a web portal and a mobile application have been developed.

V. IMPLEMENTATION ISSUES

To illustrate the functionality of the proposed platform, a prototype system was implemented and deployed on a laboratory cloud computing infrastructure. Prototype system implementation was based on the Oracle 11g SOA [16]. In particular, the following components of this Suite have been used: (i) Integrated Service Environment (ISE) for developing the EHR web services and (ii) an enterprise portal for healthcare professionals, collaborating healthcare organizations and researchers to access content.

The platform used for the generation of sample patient PHRs is Tolven ePHR as it has been considered sufficient for the purpose of our research [17]. With regard to health information from medical devices, weight, blood pressure and blood glucose measurements have been taken under consideration, which are uploaded in the patient's PHR automatically via a relevant Android application [15]. Due to unavailability of actual health and social data, a sample database has been created in MySQL along with the relevant web services in order to simulate healthcare and social care providers' systems [18]. The ABAC policies have been defined using eXtensible Access Control Markup Language (XACML) [19].

The implementation of the Analytics Service was based on Hadoop Mahout framework, which features various scalable machine-learning algorithms [20]. In particular, a certain part of the larger Mahout framework has been used, namely Taste, which is a Java framework for providing personalized recommendations [21]. Taste was used in order to develop a customized recommender system which comprises (a) a DataModel containing the certain patient features mentioned in the previous section and (b) a similarity function which determines the subset of patients that appear to be similar with regard to the features incorporated in the data model (Figure 3).

VI. CONCLUDING REMARKS

The emerging field of precision medicine is speeding ahead, promising a paradigm shift in care delivery, one that removes the need for guesswork, variable diagnoses and treatment strategies based on generalized demographics. A key enabling factor for the realization of the precision medicine approach is the utilization of data leveraged from direct and indirect sources to provide a more holistic view of an individual patient. Although, it is increasingly viewed as mainstream treatment, especially in cancer research and diagnostics, precision medicine is far from being integrated in several healthcare settings, including the Emergency Care settings. This paper presents a platform for enabling precision medicine in emergency care delivery which is based upon a next-generation EHR enhanced with advanced analytics tools. Thus, it is envisaged that the platform provides an appropriate infrastructure for laying the foundation for precision medicine to bring its critical benefits in emergency care. The platform's added value stems from the fact that it provides a rich set of data concerning a patient's health, lifestyle and genes and a tool for determining personalized treatments according to a set of features he is carrying. By introducing such a platform in emergency care delivery and in clinical practice in general, clinicians can mitigate many of the inefficiencies that currently encumber care optimization. These inefficiencies may be false positives, false negatives, unnecessary treatments, over- or under-medication and all have financial and quality-of-care ramifications. System evaluation is a task to be undertaken in the near future aiming at determining the system usability. Thus, its potential weaknesses may be revealed suggesting alterations in the system design and directions for future work.

REFERENCES

[1] "Precision Medicine: New Paradigms, Risks and Opportunities," Knowledge@Wharton – SAP , 2016. [Online]. Available: http://knowledge.wharton.upenn.edu/article/precision-medicine-new-paradigms-risks-opportunities/. [Accessed: 22-Aug-2018].

[2] L. Timmerman, "What's in a Name? A Lot, When It Comes to 'Precision Medicine' | Xconomy," Xconomy, 2013. [Online]. Available: http://www.xconomy.com/national/2013/02/04/whats-in-a-name-a-lot-when-it-comes-to-precision-medicine/. [Accessed: 22-Aug-2018].

[3] A. T. Limkakeng et al., "Systematic Molecular Phenotyping: A Path Toward Precision Emergency Medicine?," *Acad. Emerg. Med.*, vol. 23, no. 10, pp. 1097–1106, Oct. 2016.

[4] C. Shively, "Beyond the Genome and Pharmacogenetics: Precision Medicine - Ask Dr. S," 2017. [Online]. Available: http://askdrs.org/pharmacy-successhttp://askdrs.org/beyond-the-genome-and-pharmacogenetics-precision-medicinehttp://askdrs.org/116.html. [Accessed: 22-Aug-2018].

[5] "Looking Beyond Genomics: A Holistic Approach to Precision Medicine," Evolution Global. [Online]. Available: https://evobiotalent.com/news-analysis/looking-beyond-genomics-a-holistic-approach-to-precision-medicine/. [Accessed: 22-Aug-2018].

[6] V. Koufi, F. Malamateniou, A. Tsohou, and G. Vassilacopoulos, "A framework for privacy-preserving access

to next-generation EHRs.," *Stud. Health Technol. Inform.*, vol. 205, pp. 740–4, 2014.

[7] F. S. Collins and H. Varmus, "A New Initiative on Precision Medicine," *N. Engl. J. Med.*, vol. 372, no. 9, pp. 793–795, Feb. 2015.

[8] M. Poulymenopoulou, F. Malamateniou, and G. Vassilacopoulos, "Emergency Healthcare Process Automation Using Mobile Computing and Cloud Services," *J. Med. Syst.*, vol. 36, no. 5, pp. 3233–3241, Oct. 2012.

[9] K. Rhrissorrakrai, T. Koyama, and L. Parida, "Watson for Genomics: Moving Personalized Medicine Forward.," Trends in cancer, vol. 2, no. 8, pp. 392–395, Aug. 2016.

[10] "N-of-One." [Online]. Available: https://n-of-one.com/. [Accessed: 22-Aug-2018].

[11] "2bPrecise." [Online]. Available: https://2bprecisehealth.com/about-us/. [Accessed: 22-Aug-2018].

[12] C. D. Meyerhoefer et al., "Provider and patient satisfaction with the integration of ambulatory and hospital EHR systems," *J. Am. Med. Informatics Assoc.*, vol. 25, no. 8, pp. 1054–1063, Aug. 2018.

[13] T. Heart, O. Ben-Assuli, and I. Shabtai, "A review of PHR, EMR and EHR integration: A more personalized healthcare and public health policy," *Heal. Policy Technol.*, vol. 6, no. 1, pp. 20–25, Mar. 2017.

[14] F. Prasser, O. Kohlbacher, U. Mansmann, B. Bauer, and K. Kuhn, "Data Integration for Future Medicine (DIFUTURE)," *Methods Inf. Med.*, vol. 57, no. S 01, pp. e57–e65, Jul. 2018.

[15] V. Koufi, F. Malamateniou, and G. Vassilacopoulos, "An Android-enabled mobile framework for ensuring quality of life through patient-centric care.," *Stud. Health Technol. Inform.*, vol. 180, pp. 1040–4, 2012.

[16] "SOA Suite for On-Premises Integration | Oracle," Oracle. [Online]. Available: https://www.oracle.com/middleware/application-integration/products/soa-suite.html. [Accessed: 22-Aug-2018].

[17] "Tolven ePHR," Tolven Healthcare Innovations. [Online]. Available: http://www.tolven.org/ephr.html. [Accessed: 22-Aug-2018].

[18] "MySQL." [Online]. Available: https://www.mysql.com/. [Accessed: 22-Aug-2018].

[19] "OASIS eXtensible Access Control Markup Language (XACML) TC | OASIS," OASIS. [Online]. Available: https://www.oasis-open.org/committees/tc_home.php?wg_abbrev=xacml. [Accessed: 22-Aug-2018].

[20] "Apache Mahout: Scalable machine learning and data mining." [Online]. Available: http://mahout.apache.org/. [Accessed: 22-Aug-2018].

[21] "Apache Mahout - Taste Documentation." [Online]. Available: http://svn.apache.org/repos/asf/lucene/mahout/site/publish/taste.pdf?p=836600. [Accessed: 22-Aug-2018].

Enacting Eldertech in Senior Citizens' Communities of Japan: A Social Support Perspective

Tingting Huang
Graduate School of Economics and Management
Tohoku University
Sendai, Japan
Email: yoruhuang@gmail.com

Abstract—**This work-in-progress research focuses on "the ICT enabled community-based support system" in Japan, which has done a great job on improving the quality of the seniors' social life. The goal is to adopt a qualitative research approach with a multiple case design and the Social support theory to zoom in the research questions: 1) how do various communities use the deployed ICT to support the social life of the seniors? 2) How the general Critical Success Factors (CSF) acts during constructing and managing the ICT project? The originality of this research is that it takes multiple representative cases and reforms them into a generalized conceptual model for further research and application.**

Keywords-ICT; social support; senior; community; CSF.

I. Introduction

In Japan, the need to explore practical ways to maintain senior citizens' functional abilities and promote independent living is urgent [1]. The proportion of people aged 65+ years in the total population in Japan is more than 27% [2], which means that the Japanese population aging is still unprecedented in the world [1]. Meanwhile, other issues like the shrinking working population, the low birth rate, etc. will lead to the shortage of seniors supporting staff and the growth in social security costs [3]. The projections for 2050 show that considerable countries will face the similar population challenge soon [4].

Today, various assistive technologies have been adopted to help the seniors improving their living quality, such as general Information and Communication Technologies (ICTs), robotics, telemedicine, sensor technology, medication management applications, and video games [5]. On the other hand, how to use assistive technology fulfilling mental needs of the seniors is a significant research topic [6]. Loneliness and social isolation [6] is one of the most severe problems affecting the elder, which is expressed in little participation in social activities, dissatisfaction with social lives, or even no contact with other people at all.

The experiences of Japan suggested that one of the solutions to loneliness and social isolation is to rebuilding community-based support systems [1]. Close-knit community-based social networks enhanced provision and receipt of social support [1], which is fundamental to ease the critical issue—how to provide appropriate support to the seniors who live alone [7]. Nevertheless, only a limited number of studies [8][9] in Japan have focused on this topic from a theoretical perspective.

To address the above research gaps, this research aims to select the representative cases of ICT enabled community-based support system in Japan, which have done a great job on improving the quality of the seniors' social life. In order to choose a better angle to look into the research question of this research, the Social Support Theory (SST) [12] is considered to be a perfect fit for this study to zoom in the following questions: 1) how do various communities use the deployed ICT to support the social life of the seniors? 2) How the general CSF acts during constructing and managing the ICT project? In order to explore the above questions, the three key notions of SST might provide a comprehensive perspective. Using a qualitative method to analyze the multiple representative cases [13] can clearly and concisely tell the development of the appropriate designs which combine the various ICTs and the communities' activities [14][15][16]. The originality of this research is that it takes multiple representative cases in Eldertech field for the first time and reforms them into a generalized conceptual model for further research and application.

The theoretical and practical contributions of this research are expected in the following. First, there is limited research that focuses on the ICT enabled community-based support systems and what they can do for the seniors. This study can address critical issues and contributes to the growing gerotechnology research. Second, this research might extend the adopted theory to form a new framework for explaining the similar phenomenon. Third, Japan's successful experiences in developing and operating the ICT enabled community-based support systems might present examples for other countries, especially the Asian countries that will face the population issues soon. Finally, this research might offer other practitioners a specific agenda for developing the senior support system.

The outline of this paper is in the following. A literature review of the phenomenon and the theory lens will be explained In Section 2. Section 3 is about the methodology. Case description and expectable findings are in Section 4 and 5. Finally, there is the conclusion.

II. LITERATURE REVIEW

In this section, the background of the phenomenon and the social support theory will be introduced.

A. the Eldertech

There is an increasing number of studies focused on assistive technology for the seniors living with chronic illness in their own home within the medical health, nursing, and gerontology literature [5][17]. The issues related to material needs and the physical condition of the seniors population are usually adopted the telemedicine systems [3][5][11], mobile social alarming systems and online monitoring systems based on sensor technology [5][11], and other devices combined with the services offered by the emerging ubiquitous computing and intelligent home appliances [3][5]. According to the findings of Khosravi, et al. [6], within the studies on various technologies applied to alleviate social isolation, the commonly adopted technologies consist of the general ICT contained the computers and the internet, social network sites, robotics, telecare systems, etc. Most of these studies used experimental methods and survey to argue the relationship between social isolation and the new assistive technology. There was little research focused on how to use the assistive technologies—appropriate design—effectively to ease the loneliness and social isolation. It is not an issue that only matters to the person in need, but also to a healthy and sustainable society in the future.

With the proliferation and ubiquity of ICTs, in recent years, the usage of ICTs in rebuilding community-based support systems certainly brings more efficiency and variety. In Japan, under the promotion of the Japanese government, considerable regions have started various ICT enabled regional activation projects [18] which include many community-based support models for the senior citizens. Nevertheless, only a limited number of studies [8][9] in Japan have focused on this topic from a theoretical perspective, which usually adopted survey methods and concluded from the users' perspective. Also, similar research [3][10][11] in the western countries generally took the technical point of view.

B. Social Support Theory

The SST perspective is well-suited to address our research subjects. Some key concepts of this theory are constructive on knitting critical features of our research subjects together. Social support refers to the availability of interpersonal resources [19]. SST focuses on the ways individuals are embedded in a social network through social connections, and how these connections are used to request or offer support [19]. There are three key concepts in SST. First, **social embeddedness** focuses on the structure of the social network of individuals in terms of size or density and covers the actual connections among individuals in the social environment. Second, **perceived social support** focuses on an individual's beliefs about whether or not members of one's social network provide support and the positive as well as negative consequences of these beliefs and is used to explain the effects on those individuals confronted with adverse events. Third, **enacted social support** focuses on network members who perform behavioral actions when assisting a specific individual in their networks and is used to explain behavioral actions and their consequences performed by others to provide social support. This concept evaluates an individual's actual behavior when providing support and the associated consequences.

III. METHODOLOGY

A qualitative research approach is adopted with a multiple case design to find answers to the "how" research questions [20][21]. This research will reveal the nature of the deployed ICT embedded in the applying and operating support system in each case, the relationship between the case background and the chosen ICT plan, the CSF, the effect of social support and the ICT. In the field of information systems, the CSF of the adopted information systems is well studied. The CSF usually refers to the factors that determine the success of a system to a great extent [23]. The CSF in this research means the design, task, management, and procedure, which are the most critical to the success of the system. Four representative cases will be studied mainly through in-depth interviews, observations, and related documents. The interview plan is constructed through the selected theory lens. The Social support theory provides three concepts which will enrich the dimensions of the research data—not only the subjective feelings of the end users as the previous research. For instance, social embeddedness focuses on the structure of the social network of individuals regarding size or density. It provides an objective measurement; perceived social support focuses on an individual's beliefs about whether or not members of one's social network provide support; enacted social support focuses on network members who perform behavioral actions when giving assistance to a specific individual in their networks, which reveals the efforts to achieve the ideal social integration. The observations consist of the use of the support systems, the management of the support systems, the routine and exceptional operations, the issues, etc. Finally, based on the Social integration measurement and the ICT measurement, four clear ICT usage patterns from four communities and the corresponding CSF will be concluded and form a conceptual model.

IV. CASE STUDIES

Four selected cases in Japan are analyzed in this research, which had been promoted as the successful ICT usage cases from the Japanese Ministry of Internal Affairs and Communications [22]. Table 1 shows the relevant information of the four cases. In case IRO, a leaf business is called 'Irodori'. Irodori business means cultivating, processing, and selling natural plants that are used as decorations for Japanese dishes. Less than 2000 people live in Kamikatsu, and young people take only 7.8 percent of the local population. It is the investment to the information network infrastructure (not advanced technology but the technology that can be used easily by the seniors) in 2000 that made Kamikatsu Town the No.1 of the leaf business. In the OTS case, social media (e.g., facebook & twitter) combined the official website intends to draw citizens and travelers to engage in various events in Otsuki. Seniors in Otsuki can rent their farmland to visitors, manage it for them, and post the status of the plants regularly by the social media. In the TOK case, an easy-to-use application for a smartphone is developed for communicating regularly with the local seniors who live by themselves in order to ensure their safety; they can also communicate with others easily. In the KAT case, multiple information community centers with many IT/ICT devises are set up. The seniors can gather to the nearest community center to report their safety, to use various devices, and to communicate with each other.

The investigations will be conducted following the research plan (Figure 1). The research subjects of each case include the seniors, the control organization, the support agencies, and the relevant people from the society. The plans of interviews and observations for each research subject from three perspectives are listed.

TABLE I. THE SENIORS'S SOCIAL LIFE SUPPORTED BY THE ICT ENABLED COMMUNITIES

Geographic Location	ICT project	Start Date
Kamikatsu-cho, Tokushima Prefecture	IRODORI VILLAGE KAMIKATSU (IRO)	1999
Otsuki City, Yamanashi Prefecture	HALLO NATURES OTSUKI (OTS)	2012
Tokushima City, Tokushima Prefecture	TOKUTTER (TOK)	2010
Katsuragi City, Nara Prefecture	NEW ERA KATSURAGI CREATION PROMOTION PROJECT (KAT)	2013

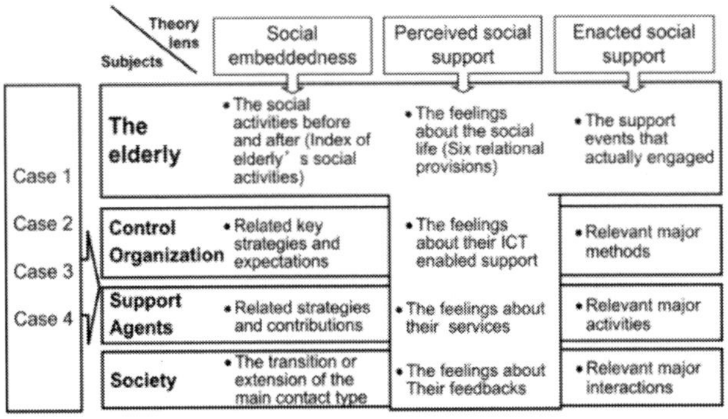

Figure 1. Investigation plan in detail.

V. EXPECTED FINDINGS

Based on the previous analysis of the secondary data of the four cases, this research intend to form a 2×2 matrix with Social integration and Social network platform dimensions (Figure 2). Social integration is a sense of belonging to a group of people who share common interests and recreational activities. Public or private means whether the platform belongs to a third-party. The characteristics of each case's findings will be discussed through the theoretical lens as well as the rational enablers. In detail, the research agenda in each case consists of the strategy and method of the control organization, the social situation and feelings of the elderly, the primary interactions with support agents and society, the CSF of the case, the distinguishing features of the design and outcomes. In the current state, the contents in Figure 2—the titles of the above research agenda—are the same; they will be specified during the investigation. For instance, the 'strategy and method of control organization' in four places of Figure 2 will be replaced by the research findings, such as 'Designing, constructing, and operating the whole information system and hardware for the seniors' in Case IRO, 'Using the social media and office website for information distribution' in Case OTS, etc.

From the current analysis, four success cases already show different features through the theory lens (Table II). Case IRO and Case TOK will be used as examples to explain the Table II. From the perspective of Social embeddedness, the seniors' social lives are analyzed through the social activities before and after. In the Case IRO, the seniors are organized by the Irodori company to engage in product manufacturing, information processing, business managing, etc. They have to interact with the company, the customers, the visitors, etc. The changes in the social activities of seniors before and after are enormous. Hence, the cell of Social embeddedness in Case IRO in Table II is filled with 'High'. In the Case TOK, each senior living alone is provided with a smartphone and the special application developed for them. The application is based on twitter and is designed for seniors to use it easily. Every day, the seniors will receive a message from support group inquiring about their status, and they should reply to this message for confirming their safety. Additionally, they can also twit something else as well. Right now, they are connected with volunteers of the support group and the other seniors in the living area, which might bring them many opportunities to attend social events and make friends. However, comparing to the seniors in the first case, the support is valued as relatively 'low'.

From the perspective of Perceived social support, the interviews for the seniors will be about their feelings regarding their current social lives, the ICT enabled support, the services, the feedback from the people of society that have interacted with them. The conjectural results are listed in Table II. The seniors in Case IRO might feel their social lives are more fulfilling than the seniors in Case TOK.

Finally, from the perspective of Enacted social support, the interviews for the seniors will be about the activities and events they experienced and the assistance they received. From the current data, in Case IRO, the seniors have engaged in learning the leaf business, learning how to use the information system and hardware to do business, information and equipment upgrade, interaction with visitors, training new trainee, etc. On the other hand, the introduction of smartphone and the application is the main support activities for the seniors in Case TOK. Hence, 'High' is for Case IRO and 'Low' is for Case TOK. Similarly, the other two cases are evaluated currently through the same process.

The ongoing investigations will provide more detail information to refine the conceptual model and to enrich the qualitative analysis. Information during constructing and managing the ICT project of each case will be analyzed for locating the CSF as well as for generating survey for the system managers. General CSF for this kind of support system can be expected.

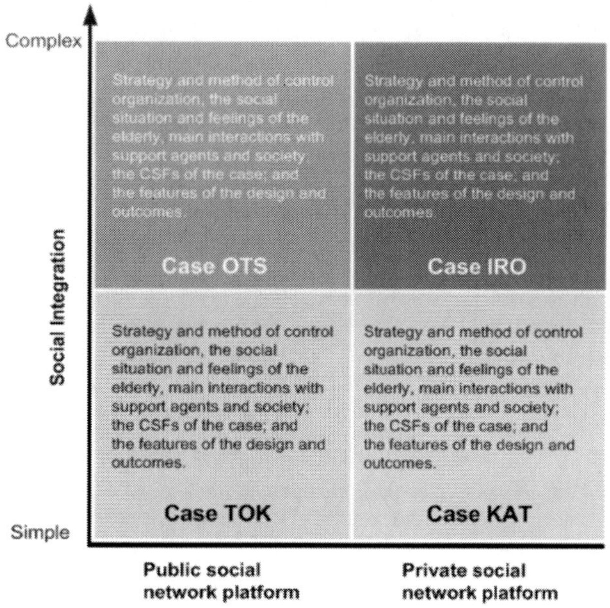

Figure 2. Different Approaches Revealed within Current Case Analyses

TABLE II. THE EXPETABLE SENIORS'S SOCIAL LIFE SUPPORTED BY THE ICT ENABLED COMMUNITIES

	ICT	Social embedded ness	Perceived social support	Enacted social support
IRO	Private	High	High	High
OTS	Public	High	High	Low
TOK	Public	Low	Low	Low
KAT	Private	Low	Low	High

*The most complex Social Integration is presented with 'High' in all three notions; the simplest Social Integration is presented with 'Low' in all three notions.

The findings are highly expected to answer the research questions. From the primary analysis of secondary data, four success cases have already shown the distinguishing features on supporting the seniors with different approaches of deployed ICT. Although the current data have not covered the identification of the CSF, the further investigations are expected to generate more valuable data on the projects' implementation and management details.

VI. CONCLUSIONS

This research focuses on "the ICT enabled community-based support system" in Japan, which has done a great job on improving the quality of the seniors' social life. There is limited research that focuses on the ICT enabled community-based support systems and what they can do for the seniors. The CSF of these success cases has not been studied well enough in the prior research as well. This study can address critical issues, such as the key issues and the CSF on ICT usage in a community, and contributes to the growing gerotechnology research. This research might extend the adopted theory to form a new framework for explaining the similar phenomenon. The previous research of ICT enabled community-based support systems usually adopted a survey method to evaluate the satisfaction of the seniors. In this research, the social support theory provides multiple perspectives not only from the seniors, but also from the control organization, the support agencies, and the society. Japan's successful experiences in developing and operating the ICT enabled community-based support systems might present examples for other countries, especially the Asian countries that will face the population issues soon. Comparing to other studies on the similar subject, the conceptual model proposed by this research will provide four success examples with specific CFS and the ICT usage patterns. Other communities can modify or refine their support systems by choosing one success example with similar background and ICT application. This research might offer other practitioners in Japan a specific agenda for developing and implementing the community-based support system and expand the research approaches on the similar phenomenon.

REFERENCES

[1] N. Muramatsu and H. Akiyama, "Japan: Super-Aging Society Preparing for the Future," The Gerontologist, vol. 51, pp. 425-432, Aug. 2011, doi: 10.1093/geront/gnr067.
[2] Statistics Bureau. *Population density*. [Online] Available from: http://www.stat.go.jp/data/jinsui/new.htm/ [retrieved: August, 2018].
[3] L. M. Camarinha-Matos and H. Afsarmanesh, "Virtual communities and seniors support," MIV'01 in Advances in Automation, Multimedia and Video Systems, and Modern Computer Science, WSES press, Sep. 2001, pp. 279-284, ISBN: 960-8052-40-8.
[4] World Health Organization, World report on ageing and health: World Health Organization, 2015.
[5] P. Khosravi and A. H. Ghapanchi, "Investigating the effectiveness of technologies applied to assist seniors: A systematic literature review," International Journal of Medical Informatics, vol. 85, pp. 17-26, Jan. 2016, doi: 10.1016/j.ijmedinf.2015.05.014.
[6] P. Khosravi, A. Rezvani, and A. Wiewiora, "The impact of technology on older adults' social isolation," Computers in Human Behavior, vol. 63, pp. 594-603, Oct. 2016, doi:

10.1016/j.chb.2016.05.092.

[7] T. Obi, D. Ishmatova, and N. Iwasaki, "Promoting ICT innovations for the ageing population in Japan," International Journal of Medical Informatics, vol. 82, pp. 47-62, Apr. 2013, doi: 10.1016/j.ijmedinf.2012.05.004.

[8] K. Kobayashi and K. Goto, "Development, Present State and New Initiatives of the "Seniors Person Watching System" : Case Studies on FUKUYAMA and FUKUOKA," Journal of Nagoya Gakuin University, vol. 52, pp. 23-38, 2016, doi: 10.15012/00000643.

[9] C. Jia and T. Noda, "Investigation on the Effect of Services for Aged Persons by Using Information and Communication Technology in Okuizumo-Cho," San'in Research, vol. 4, pp. 23-36, 2011.

[10] L. M. Camarinha-Matos and W. Vieira, "Intelligent mobile agents in seniors care," Robotics and Autonomous Systems, vol. 27, pp. 59-75, Apr. 1999, doi: 10.1016/S0921-8890(98)00083-9.

[11] L. Camarinha-Matos and H. Afsarmanesh, "TeleCARE: Collaborative virtual seniors care support communities," The Journal on Information Technology in Healthcare, vol. 2, pp. 73-86, 2004.

[12] M. Barrera, "Distinctions between social support concepts, measures, and models," American journal of community psychology, vol. 14, pp. 413-445, Aug. 1986, doi: 10.1007/BF00922627.

[13] S. L. Pan, G. Pan, and D. E. Leidner, "Crisis Response Information Networks," Journal of the Association for Information Systems, vol. 13, pp. 518-555, Jan. 2012, doi: 10.17705/1jais.00283

[14] C. L. M. Ling, S. L. Pan, P. Ractham, and L. Kaewkitipong, "ICT-enabled community empowerment in crisis response: Social media in Thailand flooding 2011," Journal of the Association for Information Systems, vol. 16, pp. 174-212, Mar. 2015, doi: 10.17705/1jais.00390.

[15] Y. Tim, S. L. Pan, S. Bahri, and A. Fauzi, "Digitally enabled affordances for community - driven environmental movement in rural Malaysia," Information Systems Journal, vol. 2017, pp. 1-28, Mar. 2017, doi:10.1111/isj.12140.

[16] Y. Tim, S. L. Pan, S. Bahri, and A. Fauzi, "Digitally enabled crime-fighting communities: Harnessing the boundary spanning competence of social media for civic engagement," Information & Management, vol. 54, pp. 177-188, Mar. 2017, doi:10.1016/j.im.2016.05.006.

[17] L. Magnusson, E. Hanson, and M. Borg, "A literature review study of information and communication technology as a support for frail older people living at home and their family carers," Technology and Disability, vol. 16, pp. 223-235, Jan. 2004.

[18] Y. Takata, "Transition of the poligy on ICT promotion in regional community," Media・Communication, vol. 62, pp. 135-147, 2012.

[19] C. Maier, S. Laumer, A. Eckhardt, and T. Weitzel, "Giving too much social support: social overload on social networking sites," European Journal of Information Systems, vol. 24, pp. 447-464, Sep. 2015, doi: 10.1057/ejis.2014.3.

[20] S. L. Pan and B. Tan, "Demystifying case research: A structured–pragmatic–situational (SPS) approach to conducting case studies," Information and Organization, vol. 21, pp. 161-176, Nov. 2011, doi:10.1016/j.infoandorg.2011.07.001.

[21] G. Walsham, "Interpretive case studies in IS research: nature and method," European Journal of information systems, vol. 4, pp. 74-81, 1995, doi:10.1057/ejis.1995.9

[22] MIT, ICT usage cases, [Online] Available from: http://www.soumu.go.jp/main_sosiki/joho_tsusin/top/local_support/ict/jirei/index.html#anc02 [retrieved: August, 2018].

[23] E. J. Umble., R. R. Haft, & M. M. Umble, "Enterprise resource planning: Implementation procedures and critical success factors," European journal of operational research, vol. 146, pp. 241-257, 2003, doi: 10.1016/S0377-2217(02)00547-7

Evaluating Real-Time Hand Gesture Recognition for Automotive Applications in Elderly Population: Cognitive Load, User Experience and Usability Degree

Diana Trojaniello, Alessia Cristiano, Alberto Sanna
Center for Advanced Technology and Wellbeing
Ospedale San Raffaele
Milan, Italy
E-mail: trojaniello.diana@hsr.it, cristiano.alessia@hsr.it, sanna.alberto@hsr.it

Stela Musteata
Division of Medical Sciences
University of Victoria
Victoria, Canada
E-mail: stella.musteata@gmail.com

Abstract— **Driving a car represents a crucial aspect to keep independence, social life and wellbeing for elderly people. Due to the age-related cognitive decline, solutions aimed to help older adults to interact easily with the vehicle and to control the car sub-systems are required. Thanks to the technology advancement, a number of interaction modalities are available, including touch, voice and, most recently, gesture control. Systems based on gesture control allow the subjects to interact with the vehicle sub-systems (i.e., vehicle navigation tool) through easy gestures, thus avoiding the subjects to be distracted while driving. This represents an interesting feature for elderly people who often show limitations in attention. On the other hand, learning the use of new technologies, such as a new interaction modality, as for the gesture control based systems, could represent a critical issue, in particular for elderly people who often suffer of memory problems. The current study aims to investigate the usability, user experience and mental workload associated with the first usage of a new developed prototype of an in-vehicle system, based on gesture control, for elderly people. Results showed that a low usability degree, as well as a quite high mental workload, is associated to the usage of the proposed prototype. The inclusion of other interaction modalities, such as voice and touch controls, as well as the improvement in the gesture control system, i.e., by reducing the number of gestures needed, is required in future releases of the developed prototype.**

Keywords- Ageing; Elderly; Mobility; HMI; Human Machine Interaction; Gesture control; Usability; User experience

I. INTRODUCTION

The European population estimated in 2017 is about 511.8 million people, and 19% are older adults aged 65 and over (Eurostat). The life expectancy increase has brought great revolutions in the social and cultural spheres, and new challenges in relation to the health and well-being of older population.

Cognitive decline and brain aging is one of the older adult's challenges. Some seniors maintain excellent cognitive functions up to 70 or 80 years, others show signs of cognitive decline already in their 60s. Attention, processing speed [1] and episodic memory [2] represent the mostly aging affected cognitive functions with dramatic consequences on the daily tasks performances, such as driving [3]–[5]. A prerequisite for driving is the integration of high-level cognitive functions with perception and motor functions. The cognitive functions involved in driving are divided attention, processing speed, visual perception, short-term memory, working memory and episodic, semantic and procedural memory [6]–[8]. The age-related decline of these abilities in older drivers leads to difficulties of handling trafficked intersections and high speed roads, noticing the nearby upstream signals, negotiating wide multi-lane carriageways, etc. [7]. For example, due to the age-related decline of cognitive abilities, many of European seniors consider driving a car a stressful task. Furthermore, the UK Department for Transport research suggests that drivers aged between 60 to 69 had in average 18.8 crashes casualties per billion miles driven [9]. This number significantly increases to 56.7 casualties for drivers older than 70 years.

Reaching the grocery shops, the neighborhood facilities, or participating in community social events constitute essential needs for an older adult; therefore, the main challenge of nowadays society is to preserve the driving ability of older population. Maintaining a good mobility is vital and highlights the importance of developing innovative solutions that will help the ageing population to feel confident in driving safely. In this context, the design of future in-vehicle interfaces should take into account older drivers' needs and capabilities by increasing the safety and the comfort of elderly people. Intelligent Transport Systems, including in-vehicle navigation systems (i.e., tools that use geographical information to give feedback and support to drivers) can provide older drivers with increased confidence, and potentially deter them from taking risky behaviors [10]. In addition, new human-machine interfaces should be more accessible without requiring long periods of learning and adaptation. They should also provide more natural human-machine interaction avoiding overloading the mental abilities of older drivers.

As described in Myron Krueger's book Artificial Reality (1993), "natural interaction" means voice and gesture [11]. The voice control of in-vehicle systems is seen as an extremely desirable feature and potentially safe application for older adults, allowing to drive without requiring visual attention [12]. Similarly, in the last years, hand gesture control has gained popularity due to the potential reduction of the visual load and visual distractions associated to its usage while driving. In fact, systems based on gesture control are able to distinguish hand movements while giving correspondent reaction and vocal feedback (i.e., answer the phone, send messages, listen the desired music) without diverting the attention from driving. Furthermore, gesture control does not limit the autonomy and safety of the driver, while reducing the errors in driving.

An innovative prototype of Human Machine Interface (HMI) system based on gesture control has been realized within the H2020 European project "SILVERSTREAM". It has been specifically designed to help elderly people while driving, avoiding unambiguous and problematic interactions.

The objective of the present study was to investigate various aspects of the proposed HMI system by assessing its suitability for the elderly people through an evaluation of user expectation, user experience and usability, as well as an assessment of the mental workload associated with its usage.

The paper is organized as follows. An overview of the system, including the hand tracking device, as well as the sample population and acquisition protocol description are reported in section II. While section III provides the experimental results, the discussion has been provided in section IV. Finally, the conclusion and possible future works are reported in section V.

II. Materials and Methods

In the following sections an overview of the system as well as of the sample population and experimental activities along with the data analysis plan has been provided.

A. *System*

The tested system consisted of the following components:

- **Hand tracking device** (Leap motion controller, Leap Motion, Inc., USA): a small USB peripheral device which use two monochromatic IR cameras and three infrared LEDs to track the hand gestures and recognize the fingers movements (Figure 1a);
- **Laptop** (Notebook F302LJ, ASUS) where the developed software for vehicle management (i.e., home, settings, radio, car navigation, etc.) has been installed (Figure 1b);
- **Monitor** (FA1013/S 10.1 inches, Lilliput) (Figure 1c);
- **3D mouse** (Space Mouse Compact, 3Dconnexion, Germany), which includes an internal 6 degrees of freedom sensor allowing to zoom and rotate in an intuitive manner thanks to simple movements. (Figure 1d).

The above-mentioned components have been hardware connected and represent the tested "HMI system".

B. *Hand tracking device*

The hand tracking device (Leap motion sensor) represents the core of the tested system since it allows the software remote control by tracking the hands and fingers movements (i.e., "gestures"). The device consists of two IR cameras and three infrared LEDs directed along the y-axis with a field of view of about 150 degrees (Figure 2).

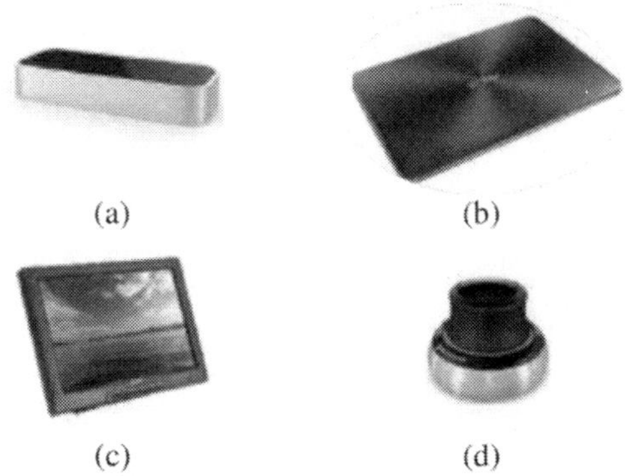

Figure 1. System components

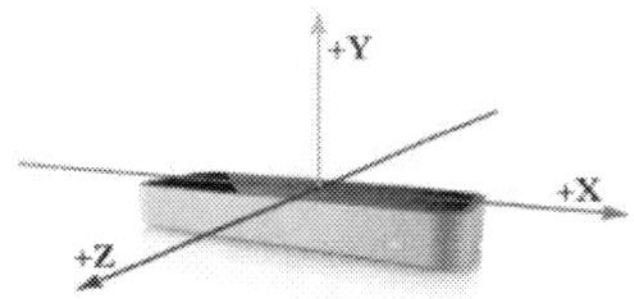

Figure 2. Leap motion sensor coordinate system

The effective range covered is from approximately 25 to 600 millimetres above the device. Position of hands and fingers is provided thanks to a model of human hand included in the linked proprietary software. The software recognises certain movement patterns (i.e., gestures), which indicate the user's intention or command. The recognized gestures could be clustered in the following subclasses:

- Circle Gesture (Figure 3a)
- Swipe Gesture (Figure 3b)
- Back Gesture (Figure 3c)
- Grab Gesture (Figure 3d)
- Hand Key Tap Gesture (Figure 3e)

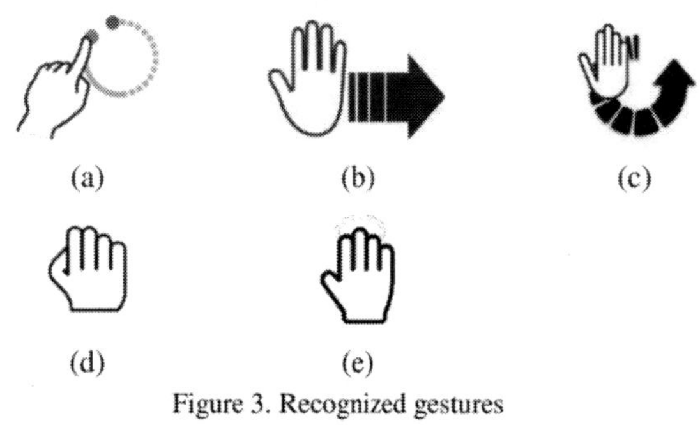

Figure 3. Recognized gestures

C. *Sample Population*

A sample of thirty subjects aged over 65 years old took part to the study. Subjects were recruited from a previous clinical study ("Epidemiological study on a sample of elderly subjects with subjective complaints of memory") approved by the Ethics Committee of San Raffaele Hospital (HSR), in Milan. All participants provided informed written consent, edited in accordance to the Declaration of Helsinki [13].

D. Experimental Setup

Subjects have been invited at the HSR facilities to take part to the experiment. A room, suitably furnished with the HMI system, has been chosen as scenario for the experiments. A large digital screen has been used during the experiments to show the gestures to the participants and the requested tasks (Figure 4).

Figure 4. Experimental setup

Two researchers, i.e., testers (a neuropsychologist and a bioengineer) were involved in the experiments.

E. Acquisition protocol

Before starting the experiment, participants have been interviewed through "Preliminary Interview" by the neuropsychologist regarding their level of confidence in using technological devices, in both general and automotive context. Then, the following experimental protocol (lasting about 1h and half) has been carried out for each participant:

- **Cognitive assessment** through the Mini Mental State Examination (MMSE);
- **Demonstrative video** illustrating the main features of the HMI system (i.e., setting, gestures, etc.);
- **Subjective expectations assessment** through the user experience evaluation questionnaire ($SUXES_i$) [14], to assess the user expectation about the proposed system;
- **Familiarization period** in which participant was invited to freely use the HMI system;
- **Usability test** where the participant was asked to carry out a number of tasks commonly performed in a vehicle (e.g. selection of an audio track through the HMI system or regulating the temperature) through the gestures shown in the demonstrative video and reported in Figure 3;
- **Subjective usability assessment** through the System Usability Scale (SUS) questionnaire [15], [16], to evaluate the post-test perceived usability of the system;
- **Subjective experience assessment** through the $SUXES_f$ questionnaire, to assess the user experience after the system usage.
- **Mental workload assessment** through the NASA [17] Task Load Index (TLX) questionnaire.

Finally, the neuropsychologist interviewed the participant by means of "Final Interview" to collect his/her impressions about the tested system focusing in particular on: ease of use, workload associated and main difficulties found.

F. Methods

All the above-mentioned surveys have been already validated and published in literature and represent gold standard methodologies for assessing the following aspects of interest: cognitive functions (MMSE), users' expectations and users experience (SUXES), the system usability (SUS) and mental workload (NASA) associated to the HMI system use.

The usability test has been performed according to the standard guidelines [18], which requires the participant to perform a number of task using the tested system. During the test, the tester observed the participant without formulate any question while collecting some quantitative variables (i.e., required time for the tasks, number of attempts and number of errors). These variables were useful for the objective evaluation of usability of the HMI system.

G. Data analysis

1) Preliminary Interview

The general knowledge of the technology along with the confidence in the usage of the three more common interaction modalities in automotive context (voice, touch and gesture) have been investigated for each subject. A frequency analysis has been performed considering the answers obtained.

2) Cognitive assessment

Each participant's MMSE outcome score was corrected for age and education and compared with the pathological cut-off of 23.60/30.

3) User expectation and user experience

According to [14], for each subject, two different scores associated to the user expectations (the "desired" and the "accepted" level) have been identified through the $SUXES_i$. Instead, a further score associated to the user experience (the "perceived" level) has been computed through the $SUXES_f$. Based on such scores, two different measures have been computed:

- Measure of Service Superiority (MSS): difference between the perceived level and the desired level;
- Measure of Service Adequacy (MSA): difference between the perceived level and the accepted level.

Those measures allowed the estimation of the gap between expectation and experience. If the experience is in the range of expectation, MSS value is negative and MSA is positive.

4) Usability of the system

According to the level of agreement selected for each statement of the SUS questionnaire, a SUS score has been computed for each participant. According to [15], [16] depending on whether the reported SUS score was greater or smaller than 68, the system was defined usable or not.

In addition, in order to evaluate the subjects performances in using the gesture control, quantitative scores (i.e., objective scores) collected during the usability test (number of errors) have been analyzed for each gesture.

5) Mental workload

An overall score of workload in a 100-point scale based on weighted average of six sub-scales (Mental Request, Physical Request, Temporal Request, Performance, Effort and Frustration) has been obtained for each subject.

III. RESULTS

In the following, the main results of the study have been provided.

A. *Sample Population*

Subjects' characteristics (age and schooling) are reported in Table I.

TABLE I. PARTICIPANTS' AGE AND SCHOOLING

Gender	#	Participants' characteristics	
		Age [years]	*Schooling [years]*
Female	17	70±4	12±3
Male	13	73±3	14±3

B. *Preliminary Interview*

Subjects' knowledge of the technology (clustered in three main categories: poor, medium and high according to the self-reporting score) has been shown in Table II.

TABLE II. KNOWLEDGE OF TECHNOLOGY

	High	Medium	Poor
Knowledge of technology	40%	33%	27%

Subjects' knowledge of the interaction modalities (touch, voice and gesture control) has been reported in Figure 5.

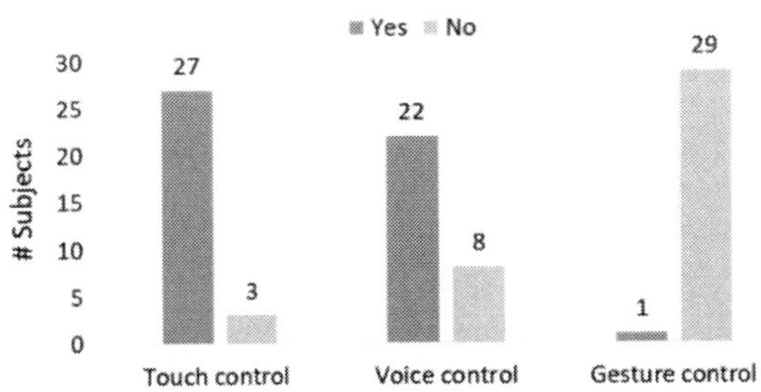

Figure 5. Knowledge of interaction modalities

Only one subject reported the knowledge of gesture control, while none of them had ever had the chance to try it.

C. *Cognitive assessment*

All the subjects reported no cognitive impairment (MMSE<23.60). The subjects reported the following scores: MMSE=26 (n=1); MMSE=27 (n=3); MMSE=28 (n=7); MMSE=29 (n=7); MMSE=30 (n=12).

D. *User expectation and user experience*

The values of MSS and MSA computed for each subject are reported in Figure 6.

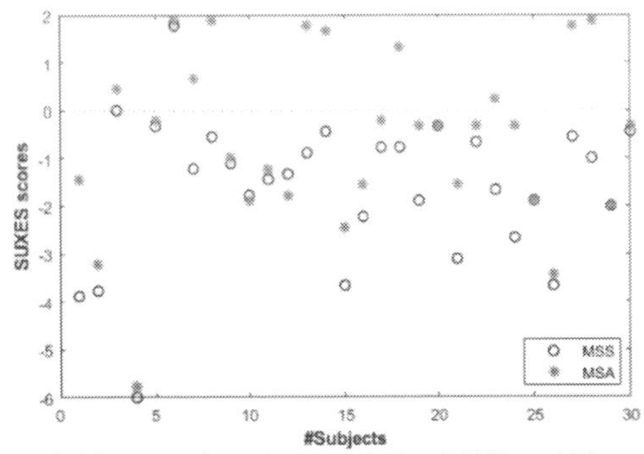

Figure 6. Measure of service superiority (MSS) and Measure of service adequacy (MSA)

As shown in the figure, for all the subjects, except two, the perceived level was lower than the desired one (MSS<0) but, for ten of them, higher than the accepted one (MSA>0). The experiences resulted to be in the range of expectations (MSS values negative and MSA values positive) for eight subjects.

E. *Usability of the HMI system*

The SUS scores are shown in Table III.

TABLE III. SUS SCORES

SUS score	Participants
> 68	27 %
≤ 68	73 %

According to the table, only 27% of subjects reported a SUS score >68 meaning that the system was judged usable. The Acceptability level has been reported in Table IV.

TABLE IV. LEVEL OF ACCEPTABILITY ACCORDING TO THE SUS SCORE

Acceptability	Range	#Subjects
Not acceptable	0-50	11
Marginal-low	50-62	9
Marginal-high	62-70	2
Acceptable	70-100	7

The usability of the system has been also analyzed considering the number of errors reported for each gesture during the usability test, as reported in Table V.

TABLE V. NUMBER OF ERRORS DURING THE USABILITY TEST

Gesture	Requested repetitions (#)	Errors (#)	
		mean	std
Circle Gesture	60	3	± 9
Swipe	600	4	± 5
Back	**180**	**8**	**± 10**
Grab	60	1	± 2
Hand Key Tap	390	4	± 5

The higher number of repetitions for some of the gestures (Swipe, Hand Key Tap and Back) are related to the tasks flow (i.e., made a phone call, select one music track, etc.).

F. *Mental workload*

The overall workload has been computed for each subject and reported in Figure 7 together with a box and whisker to graphically summarize the data.

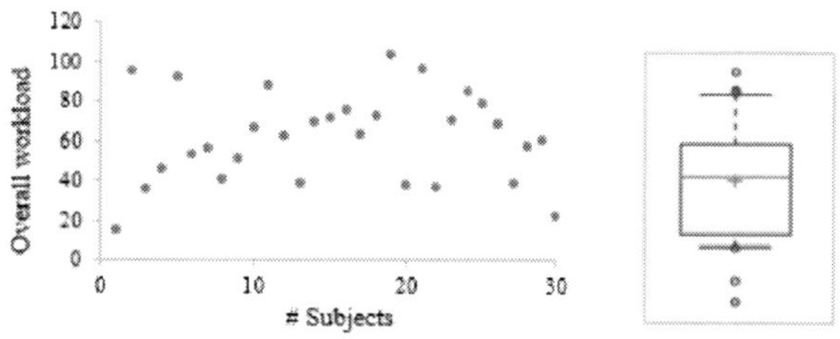

Figure 7. NASA TLX score

The workload experienced by participants for each dimension investigated by NASA TLX questionnaire has been shown in Figure 8.

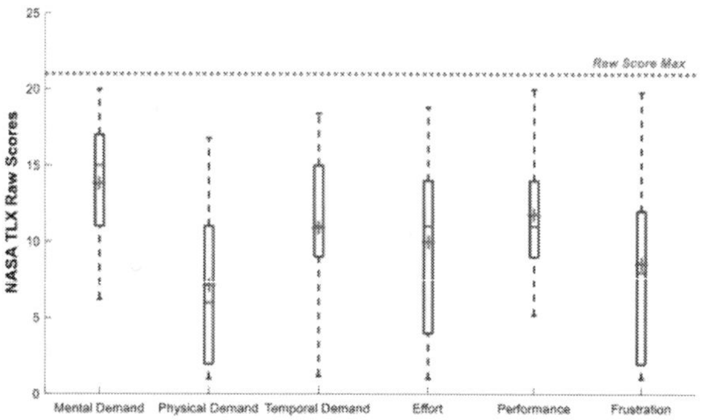

Figure 8. NASA TLX raw scores

The NASA dimension characterized by a higher workload resulted to be the mental one.

G. *Final Interview*

Subjects' impressions about the ease of use of the HMI system, the associated workload and the difficulties found during its use have been reported in Table VI.

TABLE VI. PARTICIPANTS' FEEDBACK

	Ease of use	Workload		Difficulty	
	Easy	*Physical*	*Cognitive*	*Memory*	*Association*
Yes	13 %	23 %	63 %	23 %	20 %
No	87 %	77 %	37 %	77 %	80 %

Subjects reported a low perceived ease of use (87%) and cognitive effort has been noticed (63%). Instead, only few subjects reported high physical workload (23%) and memory (23%) or association (20%) difficulties.

IV. DISCUSSIONS

The purpose of this work was to verify the appropriateness in terms of user experience and usability of the proposed HMI system to the elderly specific needs.
All subjects reported good cognitive performances on MMSE test. Most of them (73%) considered the HMI system not usable and their experience was below the range of expectations while the 63% of participants complained the cognitive workload needed to accomplish the tasks required during the test. Moreover, most participants (87%) reported that the HMI system was very difficult to learn even if only a small part of them declared, at the end of the test, to have experienced memory (23%) and association problems (20%). It is clear that such problems can lead to a lower accuracy of gestures execution, which consequently appears in negative emotions, as irritation and frustration, during the interaction with the system.

The advantage of using a gesture control-based system in automotive context does not seem to be confirmed by participants: indeed, the gestures have not been so assimilated to allow their execution without looking at the screen. A proper learning appears therefore fundamental to ensure a better experience with the proposed system and it can be done simplifying the most critical gestures as the back one, as resulted from the usability testing and, at the same time, providing more time to make the elderly more familiar with this new kind of technology.

V. CONCLUSIONS

The results obtained in the present study suggested that the proposed HMI system, based on gesture control, is difficult and not well perceived by older population. In addition, long time to ensure a correct and complete learning for properly using the system resulted to be needed.

However, it is important to highlight that the present generation of elderly has not so much familiarity with the technology contrary to the future one. Therefore, the integration with the best-known touch and voice controls, as well as an accomplished learning leading to an automatic execution of gestures should be provided to make the tested HMI system a useful tool for the forthcoming generations of elderly.

Acknowledgment

The research leading to these results has received funding from the European Union's Horizon 2020 research and innovation programme under grant agreement No 653861 – SILVERSTREAM.

References

[1] T. a Salthouse, "Aging associations: influence of speed on adult age differences in associative learning.," *J. Exp. Psychol. Learn. Mem. Cogn.*, vol. 20, no. 2, pp. 1486–1503, 1994.

[2] X. Y. Li *et al.*, "Age- and brain region-specific changes of glucose metabolic disorder, learning, and memory dysfunction in early alzheimer's disease assessed in APP/PS1 transgenic mice using18F-FDG-PET," *Int. J. Mol. Sci.*, vol. 17, no. 10, 2016.

[3] C. Bottari, C. Dassa, C. Rainville, and É. Dutil, "The criterion-related validity of the IADL Profile with measures of executive functions, indices of trauma severity and sociodemographic characteristics," *Brain Inj.*, vol. 23, no. 4, pp. 322–335, 2009.

[4] L. C. McGuire, E. S. Ford, and U. A. Ajani, "The impact of cognitive functioning on mortality and the development of functional disability in older adults with diabetes: The second longitudinal study on aging," *BMC Geriatr.*, vol. 6, pp. 1–7, 2006.

[5] P. Missotten *et al.*, "Relationship between quality of life and cognitive decline in dementia," *Dement. Geriatr. Cogn. Disord.*, vol. 25, no. 6, pp. 564–572, 2008.

[6] S. Bao and L. N. Boyle, "Age-related differences in visual scanning at median-divided highway intersections in rural areas," *Accid. Anal. Prev.*, vol. 41, no. 1, pp. 146–152, 2009.

[7] D. R. Mayhew, H. M. Simpson, and S. A. Ferguson, "Collisions involving senior drivers: High-risk conditions and locations," *Traffic Inj. Prev.*, vol. 7, no. 2, pp. 117–124, 2006.

[8] J. S. R. Leversen, B. Hopkins, and H. Sigmundsson, "Ageing and driving: Examining the effects of visual processing demands," *Transp. Res. Part F Traffic Psychol. Behav.*, vol. 17, pp. 1–4, 2013.

[9] N. Korner-Bitensky, A. Kua, C. von Zweck, and K. Van Benthem, "Older driver retraining: An updated systematic review of evidence of effectiveness," *J. Safety Res.*, vol. 40, no. 2, pp. 105–111, 2009.

[10] C. Emmerson, W. Guo, P. Blethe, A. Namdeo, and S. Edwards, "Fork in the road: In-vehicle navigation systems and older drivers," *Transp. Res. Part F Traffic Psychol. Behav.*, vol. 21, pp. 173–180, 2013.

[11] M. W. Krueger, "An Easy Entry Artificial Reality," 1993.

[12] J. E. Alvarez, I., López-De-Ipiña, M. K., & Gilbert, "The voice user help, a smart vehicle assistant for the elderly," *Lect. Notes Comput. Sci. (including Subser. Lect. Notes Artif. Intell. Lect. Notes Bioinformatics).*, 2012.

[13] World Medical Association, "World Medical Association Declaration of Helsinki," *Bull. world Heal. Organ.*, vol. 79, no. 4, pp. 373–374, 2001.

[14] M. Turunen, J. Hakulinen, a Melto, T. Heimonen, T. Laivo, and J. S. Hella, "- User Experience Evaluation Method for Spoken and Multimodal Interaction," *Proc.*, no. January 2016, 2009.

[15] A. Bangor, P. Kortum, and J. Miller, "Determining what individual SUS scores mean: Adding an adjective rating scale," *J. usability Stud.*, vol. 4, no. 3, pp. 114–123, 2009.

[16] J. Brooke, "SUS-A Quick and Dirty Usability Scale."

[17] Nasa, "NASA Task Load Index," *Hum. Ment. Workload*, vol. 1, no. 6, pp. 21–21, 2006.

[18] J. Nielsen and T. K. Landauer, "A mathematical model of the finding of usability problems," *Proc. SIGCHI Conf. Hum. factors Comput. Syst. - CHI '93*, pp. 206–213, 1993.

Metrics for Monitoring Patients Progress in a Rehabilitation Context

A Case Study based on Wearable Inertial Sensors

Brendan O'Flynn, Nektaria Dagini, Sonia Bottone, Andrea Urru, Salvatore Tedesco
Tyndall National Institute, University College Cork
Lee Maltings, Prospect Row, Cork, T12R5CP, Ireland
email: brendan.oflynn / nektaria.dagini / sonia.bottone / andrea.urru / salvatore.tedesco@tyndall.ie

***Abstract*— Inertial data can represent a rich source of clinically relevant information which can provide details on motor assessment in subjects undertaking a rehabilitation process. Indeed, in clinical and sport settings, motor assessment is generally conducted through simple subjective measures such as a visual assessment or questionnaire given by caregivers. As part of a mobile health application, wireless sensors such as inertial measurement units and associated data sets can help provide an objective and empirical measure of a patient's progress through rehabilitation using on body sensors. In this publication, several metrics in different domains have been considered and extrapolated from the 3D accelerometer and angular rate data sets collected on an impaired subject with knee injury, via a wearable sensing system developed at the Tyndall National Institute. These data sets were collected for different activities performed across a number of sessions as the subject progressed through the rehabilitation process. Using these data sets, a novel and effective method has been investigated in order to define a single score indicator which can provide accurate quantitative analysis of the improvement of the subject throughout their rehabilitation. The indicator compares impaired and unimpaired limb motor performance. The present work proves that the defined score indicator can be taken into account by clinicians to study the overall patients' condition and provide accurate clinical feedback as to their rehabilitative progress.**

Keywords- ACL; IMU; Inertial Sensors; Metrics; Motor Assessment; Rehabilitation; Wearables.

I. Introduction

Motor assessment is the part of biomechanics which studies the process by which the musculoskeletal system can create and control coordinated movements [1]. Voluntary movement requires the transmission of a message from the brain to the appropriate muscle which also controls the smoothness and coordination of the movement. If motor function is intact, muscles can be moved to command allowing symmetrical movements with significant strength levels. With particular reference to the treatment of patients with lower extremity injuries, literature has recently shown a paradigm shift, going from time-dependent concepts to function-based concepts [2], where qualitative and quantitative tests comparing affected and unaffected sides must be met before successfully accessing the following rehabilitation stage.

Qualitative and quantitative motor assessment is typically divided into clinimetrics, balance analysis, and gait analysis. Indexes, rating scales, questionnaires, and observational forms represent the clinical standard for knee joint assessment, including, for instance, Knee Injury and Osteoarthritis Outcome Score (KOOS), Oxford Knee Score (OKS), Tegner Lysholm Knee Scoring Scale, International Knee Documentation Committee (IKDC), Western Ontario & McMaster Universities Osteoarthritis Index (WOMAC) [3]. However, these tools are subjective and, even when utilised by experienced clinicians, may not be adequate or sensitive enough.

Gold-standard technology adopted in gait analysis for quantitative movement analysis may include camera-based motion analysis, instrumented treadmills, force platforms [4], but their application is constrained by costs, access to specialist motion labs, as well as practicality of application for larger patient/subject groups.

A viable alternative is represented by the adoption of small-size low-cost wearable sensing units whose consideration for lower-limbs monitoring during rehabilitation, in order to provide objective performance of impaired subjects throughout the process, has been growing lately. Indeed, inertial sensors, typically including accelerometers, gyroscopes, and magnetometers, have been used to derive gait parameters efficiently both in healthy and symptomatic subjects [5].

This paper describes a long-term investigation of post-injury rehabilitation carried out by using a wearable inertial system developed at the Tyndall National Institute, consisting of two sensors per limb, able to provide a complete biomechanics assessment for a series of scripted activities. The work is organized as follows. Hardware platform description and test protocol are described in Section II and III, respectively. The features extracted are illustrated in Section IV. The mathematical model is outlined in Section V. The obtained results are shown in Section VI. Finally, conclusions are drawn in the final section.

II. Hardware Platform

The Tyndall biomechanical monitoring system consists of two Tyndall Wireless Inertial Measurement Units (WIMUs) per leg [6]-[9]. The platform measures 44 × 30 × 8 mm and 7.2 g without battery as shown in Figure 1.

The WIMU is equipped with a high-performance low-power ARM Cortex-M4 32-bit microprocessor operating at a frequency up to 168 MHz part of the STM32F0407 family produced by STMicroelectronics. It also features a floating point unit single precision, high-speed embedded memories (1 Mb of Flash memory, 192 + 4 Kb of SRAM), an extensive range of enhanced I/Os and peripherals, and standard and advanced communication interfaces.

Figure 1. Tyndall Wireless Inertial Measurement Unit (WIMU).

Inertial sensors (3D accelerometer and gyroscope, MPU-9250 from Invensense) are the main sensing components on the platform and are wired to the microcontroller through the I2C communication. Sensor data can be transmitted wirelessly via a communication BLE-complaint module (Broadcom BCM20737S), with integrated ARM CM3 microcontroller unit, radio frequency and embedded Bluetooth Smart Stack, or logged to a removable Micro SD card with sampling rate of 250 Hz.

For measurement of inertial data, the Invensense MPU-9250 was chosen for its low power consumption and the high range (16g for accelerometer and 2000 deg/s for the gyroscope) with limited noise levels.

The platform also features a USB connector, battery charger, fuel gauge, external I/O connectors, three LEDs, and power switch. Even if not considered in the presented investigation, the platform could also provide additional sensing data, such as magnetic field (from the MPU-9250) and environmental data (pressure, humidity, temperature) from the Bosch BME280. All the components fit with mobile applications requirements and, averagely, the overall power consumption in TX/RX mode is 100 mA, dropping to 40 mA (17 mA) for stand-by (sleep mode).

III. Protocol for Data Collection

In conjunction with clinical partners, an experimental protocol for data collection was developed to evaluate patient progress. The rehabilitation exercises considered are walking, half-squat, hamstring curl, and flexion-extension, defined by physiotherapists as indicators of rehabilitation.

These are described as follows:

- In the walking exercise, the subject walks on a calibrated treadmill, which is operated at defined speeds (3-4-6 km/h) for approximately one minute per test.
- In the half-squat exercise, the subject stands with their feet shoulder's distance apart and arms crossed on the chest. Keeping the chest lifted, the hips are lowered about 10 inches, planting the weight in the heels. The body is then brought back up to standing by pushing through the heels.
- In the hamstring curl exercise, the subject stands and bends their knee, raising the heel toward the ceiling as far as possible without pain, relaxing the leg after each repetition. This is repeated on both legs.
- In the flexion-extension exercise, the subject lies supine on the floor and bends their knee raising it toward the chest as far as possible without pain, relaxing the leg after each repetition. This is repeated on both legs.

The system has been tested with an impaired subject. The impaired subject is a female athlete, age: 44, height: 161 cm, and weight: 52 kg, with good general health status, with a history of knee injuries and surgery (reconstructed anterior cruciate ligament in the left leg following a sporting injury). The tests were carried out during the course of the rehabilitation program, e.g., starting 1 month before surgery and finishing 7 months after surgery. Overall, the subject has been evaluated in 8 sessions through three periods: once in pre-surgery conditions (e.g., 1 month before surgery), then 6 times in a range of 20 weeks starting one month after surgery (namely short-term post-surgery), and finally once 3 months after the last data capture (e.g., during long-term post-surgery period).

A number of repetitions has been collected for each exercise, so as to provide an accurate picture of the overall conditions, and each exercise was evaluated during the majority of the data captures. Hamstring curl, as well as walking at 3 and 4 km/h, were performed at every session. Similarly, flexion-extension was always recorded except in the pre-surgery session due to subject's impairment of movement. For the same reason, half-squat and walking at 6 km/h were not recorded in the first 2 sessions after surgery.

IV. Features Extraction

The metrics considered for the patient's assessment are divided into seven categories described below. More details on the computation of the features are reported in [6]-[8].

A. Gait Metrics

Well-known gait measures are calculated from the data recorded by the inertial sensors attached on the shanks, including: Gait Cycle Time (GCT), Stance Phase (StP), Swing Phase (SwP), Stride Length (SL), Stride Speed (SS), Stride Clearance (SC). This information is obtained for both legs only for walking. This category includes 6 features.

B. Range of Motion (RoM) Metric

Knee Range of Motion (RoM), defined as the peak-to-peak amplitude of the knee joint angle over the x-, y-, and z-axis during a single exercise repetition, is obtained for both limbs and for all the exercises taken into account. This category includes 3 features.

C. Kinematic Metrics

Kinematic metrics, which have been occasionally adopted for gait analysis, can provide useful information on the movement analysis. Those metrics include: Range of Angular Velocity (RAV), Vertical Acceleration (VA), Vertical Velocity (VV), Fluency (along the three axis), Kinetic Value (KV). All those features are calculated for each of the 4 sensors used for data collection and for all the exercises. This category includes 7 features.

D. Stability Metric

Stability is defined as the dynamic time warping of the x-, y-, z-axis of the acceleration and angular rate signals measured at two consecutive repetitions/strides, then averaged based on all the repetitions present in a test session. Those features are calculated for each of the 4 sensors used for data collection and for all the exercises. This category includes 3 features.

E. Jerk-based Metrics

Jerk is the rate of change of the acceleration in a repetition. Several jerk-based metrics have been investigated in literature, including: Integrated Squared Jerk (ISJ), Mean Squared Jerk (MSJ), Cumulative Square Jerk (CSJ), Root Mean Square Jerk (RMSJ), Mean Square Jerk normalized by peak speed (N_MSJ), Integrated Absolute Jerk (IAJ), Mean Absolute Jerk normalized by peak speed (N_MAJ), Dimensionless Square Jerk (DSJ). Those features are calculated on the three axis, for each of the 4 sensors used for data collection, and for all the exercises. This category includes 24 features.

F. Statistical Metrics

This category takes into account various well-known statistical features extrapolated from the time-domain. Those variables are applied on every segmented walking stride/exercise repetition for both legs performed during the sessions. The selected features are described below:

- Mean, standard deviation, skewness, kurtosis, root mean square, calculated over the acceleration and angular velocity magnitudes,
- Mean, standard deviation, skewness, kurtosis, root mean square, minimum, maximum, Coefficient of Variation (CV), and Peak-to-Peak (p-p) amplitude over the x-, y-, and z-axis of the acceleration and angular rate signals.

All those features are calculated for each of the 4 sensors used for data collection. This category includes 64 features.

G. Spectral/Entropy/Information-Theoretic Metrics

This category takes into account various well-known spectral, entropy, and information-theoretic feature. Spectral metrics are obtained using the Fast Fourier Transform (FFT). These variables are applied on the raw 3-axis of the accelerometer/angular rates data collected for both legs on each session. All these features are calculated for each of the 4 sensors used for data collection. This category includes 74 features. The features are: Dominant Frequency (DF) and its Width (FWHM), Spectral Centroid (SpC), Power in 1.5-3 Hz (LFP), Power in 5-8 Hz (MFP), 25-50-75% Quartile Frequency (QF), Spectral Edge Frequency (SEF) at 95% (calculated on the magnitude signal), Harmonic Ratio (HR), Ratio High-Low bands (RHL), Frequency-Domain Entropy (FER), Lempel-Ziv Complexity (LZC).

While LZC is calculated on the single repetitions/strides, the other features are not extrapolated for each segmented walking strides/repetitions but are obtained for a sliding window covering the 50% of the whole signal, with 10% overlapping. The data analysis is implemented off-line over the data collected using a commercial software package (MATLAB R2015a, The MathWorks Inc., Natick, MA, 2015). Each repetition/stride was visually segmented.

V. Score Modeling

Preliminary analysis described in [6]-[8] have highlighted that several parameters are seen to be potentially relevant to provide indications on patient's performance during rehabilitation. However, to support clinicians during their clinical practice, it is essential to obtain a single indicator scoring regarding patient's performance, so as to avoid analyzing all the parameters separately.

The Mahalanobis distance is typically adopted to describe how much a patient's performance deviates from the control group. However, this distance does not allow the comparison between the data distribution related to the affected and unaffected side, which is essential for rehabilitation. Moreover, this distance assumes that control and patient group have comparable standard deviations, which cannot be assumed in patients following orthopedic injuries. A more reliable extension of this metric is the Bhattacharyya Distance D_B, which measures the similarity of two discrete distributions. D_B obtained as follows:

- Given a specific session and a specific exercise, every feature listed in Section IV is extrapolated from the raw inertial data of all 4 sensors.
- Potential outliers in the feature distribution are then detected and replaced via Winsorization.
- Once the outliers are replaced, the feature vectors for the left and right leg are considered as input of the D_B calculation with their associated averages and standard deviations. This process is repeated for every exercise, session and feature. As a result, after M sessions, for every feature the distance vector C is obtained.

An accurate assessment of a patient's performance requires the selection of the informative features from every category, excluding those uninformative or redundant. Some features can be informative for some exercises and being redundant for others; thus, it is important to define an automatic method for selecting those features.

A common technique for feature selection is the Least

Absolute Shrinkage and Selection Operator (LASSO) [10]. This regression tool requires to define an output in order to adjust the weights of a linear model which defines the features to be selected. As shown in [10], this output was defined as linearly increasing from the first to the last test session, with this period ranging from 4 to 12 days in the experiments carried out by the authors. However, even though this assumption can be accepted for the short period of time immediately following surgery, it may be unrealistic when analyzing rehabilitation outcomes for a longer period post-surgery and also pre-surgery.

An alternative feature selection approach recently studied is the Clustering Coefficients of Variation (CCV) [11], which is a light-weight and efficient method based on feature variability. The features are clustered according to their CV, and then the optimal cluster of features for the model is chosen. Features showing the most variation between limbs and between different sessions over the course of the rehabilitation represent informative features to be chosen. Therefore, for each M-dimensional distance vector *C* calculated, the associated CV is obtained. If the CV is lower than one, than the associated distance vector is discarded. Following this initial selection, the remaining distance vectors are normalized through the standard score approach and then considered as points in an M-dimensional space where they are clustered via a weighted K-means clustering (K = 2 [11]). The normalization step before the clustering is important in order to guarantee that different scaling between the features could not impact the clustering.

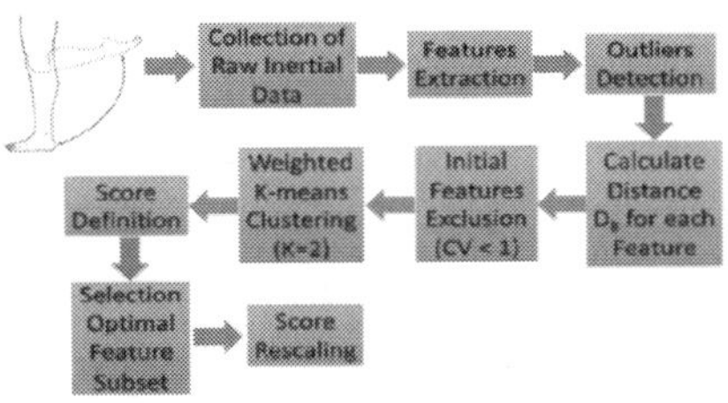

Figure 2. Scoring algorithm.

As a result of the weighted K-means, the features are divided in two clusters. For both clusters, the normalized distance vector *C* are averaged among all the features, resulting in two M-dimensional scoring vectors. These scoring vectors are then rescaled so as to be within the range [0-1]. A high score indicates a large distance in performance between limbs, and *vice versa* a low score represents a small difference. One of the two clusters (and associated scoring vectors) provides the optimal feature subset, and this selection is realized by using Hyper-Pipes [11].

To the best of the authors' knowledge, this is the first time that a combination of Bhattacharrya distance and CV-based weighted K-means clustering is investigated for monitoring patients' progress in a rehabilitation context. A summary of the scoring algorithm used is illustrated in Figure 2.

VI. Results

In each session, each exercise was divided in two separate tests (both logged for 60 sec), and in each of the two tests a series of repetitions have been carried out by the subject. The overall number of repetitions recorded for all the sessions was: 184 hamstring curls (92 left / 92 right), 134 flexion/extensions (67 left / 67 right), 66 half squats, 478 strides for both legs when walking at 3 km/h, and similarly 544 strides when walking at 4 km/h, and 512 strides when walking at 6 km/h.

WIMUs have been attached to the anterior tibia, 10 cm below the tibial tuberosity, and to the lateral thigh, 15 cm above the tibial tuberosity using surgical adhesive tape. For each test, the features, separated for every category as described in Section IV, were extrapolated and compared among the different sessions after applying the scoring method defined in Section V.

Finally, in order to have the same reference system for both WIMUs worn on the same leg, the method proposed by Seel et al. [12] has been adopted to virtually rotate around an axis the raw inertial data recorded on the shank. As a result, for all the WIMUs involved, the x-axis represents the mediolateral axis, the y-axis is the anteroposterior one, while the z-axis is the vertical axis. Thus, the plane y-z represents the sagittal plane.

Results for the metrics associated with gait, RoM and kinematics are described in the following subsections. Results for additional metrics are still under analysis and will be described in future works.

A. Gait Metrics

Considering the gait results at 3 km/h, the metrics which are clustered and show patients' progress during rehabilitation are SwP and StP. The resulting score shows a clear increase in association to the second session (due to the early stage of the recovery process post-surgery), with a consecutive decreasing trend converging toward the score value reported in the pre-surgery session. A similar behavior is also shown in the gait exercise walking at 4 km/h, also including GCT in the selected features.

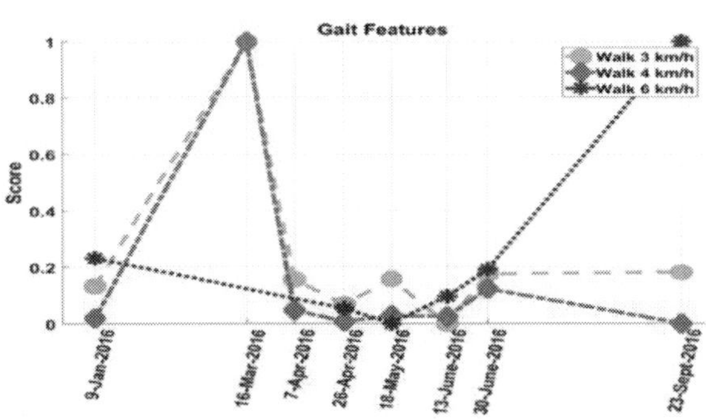

Figure 3. Score indicator. Gait features

The score highlights a peak in the post-surgery early stage with an evident convergence towards zero in the following sessions. Finally, SC is the only feature selected for the gait at 6 km/h. The score calculated for the pre/post-surgery is comparable (although this exercise was not recorded in the first two sessions after surgery because of patient's impairment) even though an unexpectedly large value is obtained in the long-term session as shown in Figure 3.

B. RoM Metrics

In the hamstring curl exercise, the mediolateral RoM (e.g., over the x-axis) shows a clear trend with a score steadily decreasing following surgery. The RoM over the z-axis is, instead, selected for the flexion/extension exercise. Even though there is a general tendency of the score to decrease starting from the second sessions, two large scores are obtained for the 6th and 8th session, indicating a non-monotonic improvement. Similar considerations can be also drawn for the squat exercise (z-axis RoM), that is a general reduction of the calculated score with an exception reported in the 7th session.

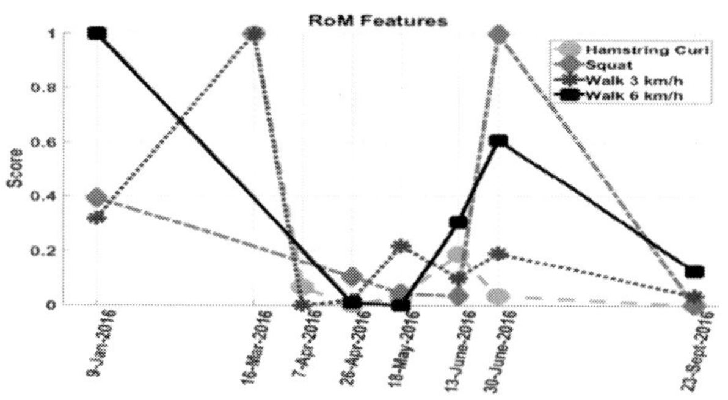

Figure 4. Score indicator. RoM features.

Walking tests at 3 and 4 km/h have both selected the x-axis RoM. In the former case, the score has a large difference from the first two sessions, while generally decreasing to low score values in the following sessions, with results lower than the pre-surgery period. In the latter, instead, the RAV score is also showing a decreasing trend with an exception in the 7th session. Finally, anteroposterior axis RoM is the feature selected for the gait at 6 km/h. The score calculated for the pre-surgery session has a much larger value in comparison with the remaining session following surgery, even though the trend in this period is clearly not monotonic. The discussed results are shown in Figure 4.

C. Kinematics Metrics

In the hamstring curl exercise, the score obtained from the features selected considering the sensor attached on the shank (e.g., VV and Fluency over the x-, and z-axis) shows a clear trend decreasing following surgery, even though the tendency in the long-term is not monotonic. Identical considerations can be drawn for the thigh sensor (with chosen features being RAV, VV, and y-axis Fluency) despite an even less flat trend in the late session; indeed, it is evident a large score value on the 6th session.

The flexion-extension exercise is described by similar conclusions, but the metrics taken into account are VV and z-axis Fluency from the shank, and only z-axis Fluency from the thigh.

Regarding the squat exercise, again the score obtained by considering VA and z-axis Fluency from the shank generally shows a decreasing trend throughout the sessions. However, when observing the score extrapolated from the thigh sensor data, no particular correlation is evident due to several large values, indicating that the thigh sensor placement is not beneficial when analyzing squats.

For walking at 3 km/h RAV and y-axis Fluency, and KV are selected for the shank and thigh, respectively. The score trend is comparable for both sensor locations, with a score presenting a large difference between pre-surgery and immediate post-surgery, and with a decreasing score reaching its minimum at the 6th session. However, large scores are shown again for the last two data collections.

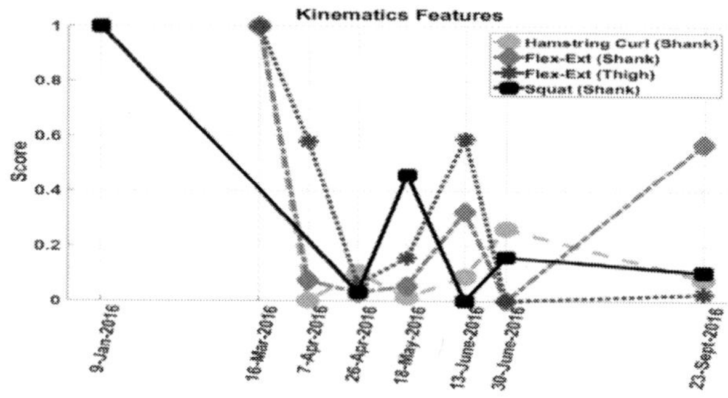

Figure 5. Score indicator. Kinematics features.

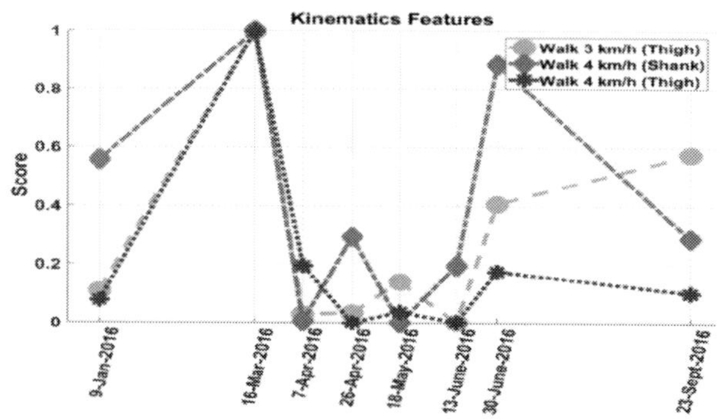

Figure 6. Score indicator. Kinematics features.

Walking at 4 km/h is not significantly different, even though in this case the selected features are RAV, x-axis Fluency, and KV for the shank. The only dissimilarities compared to the 3km/h exercise are evident in the larger score on the shank in the pre-surgery session and the flatness shown in the thigh-related score.

Finally, RAV and z-axis Fluency are the feature selected from the shank (thigh) for the walking exercise at 6 km/h. The score has a general decreasing trend when considering the shank data, even though two large values are shown on the 6th and 7th session. The data from the thigh, instead, show a score comparable between the pre-surgery and the long-term period, without any particular trend. Thus, the thigh sensor placement may be not beneficial when analyzing faster speeds. The discussed results are shown in Figures 5-6.

VII. Results Summary and Conclusions

The focus of this work-in-progress has been on the analysis of certain metrics associated with gait analysis in an effort to develop a score-based system to aid clinicians in the diagnosis and evaluation of gait in a rehabilitative context. To summarize, this work analyzed the body-worn inertial data collected from a patient over the course of rehabilitation defining a score metric from a number of features for better understanding and monitoring patient's progress and limbs comparison in several tests. It has been also shown that several metrics, gait, joint angle-related, and kinematics variables, obtained from acceleration and angular rate of the shank and thigh have proved their sensitivity for a number of exercises.

This work presented a wearable inertial system for an objective assessment of lower-limbs in patients over the course of. The hardware platform adopted for the system realization and the data analytics involving inertial data collected from thighs and shanks have been described. The present study proved that a novel scoring method involving Bhattacharyya distance metrics and Clustering Coefficient of Variation for feature selection, based on a number of well-known metrics extrapolated from inertial data collected on the lower-limbs, can be used for defining quantitatively patients' progress when involved in a rehabilitation program. Accurate results have been shown in a number of exercises. The proposed method is able to indicate which features are more informative regarding patients' performance and group them in a single indicator which can be easily taken into account by clinicians during their analysis. This score indicator represents an important step towards the development of an objective model for patients' assessment during rehabilitation.

As only a single subject has been analyzed for the present study, an enhanced number of athletes, with homogeneous characteristics, will also be tested to have a more robust base and further validate the drawn conclusions. Results associated with additional metrics, such as jerk, statistical, spectral features, are currently under investigation and will help assist in the development of such a score-based system as is envisaged by the authors. Additional clinical trials are currently being planned to further validate the developed model in statistical terms.

Acknowledgment

This publication has emanated from research supported by a research grant from Enterprise Ireland (EI) under grant number CF-2017-0685-P. Aspects of this work have emanated in part from research supported in part by Science Foundation Ireland (SFI) and is co-funded under the European Regional Development Fund, Grant Number 13/RC/2077-CONNECT and SFI-16/RC/3918-CONFIRM.

References

[1] P. D. Hansen and L. S. Chandler, "Assessment of sensory and motor function," in Rehabilitation, Human Brain Function (Assessment and Rehabilitation), G. Goldstein, S. R. Beers, Springer, Boston, MA, USA, 1998, pp. 131-153.

[2] M. Keller, E. Kurz, O. Schmidtlein, G. Welsch, and C. Anders, "Interdisciplinary assessment criteria for rehabilitation after injuries of the lower extremity: A function-based return to activity program," Sportverletz Sportshaden., 30, 1, 38-49, pp. 2016.

[3] E. C. Rodriguez-Merchan, "Knee instruments and rating scales designed to measure outcomes," J Orthop Traumatol, 13, 1, 2012.

[4] G. Robertson, G. Caldwell, J. Hamill, G. Kamen, and S. Whittlesey, Research Methods in Biomechanics, Second Edition, Human Kinetics, Champaign, IL, USA: 2014.

[5] C. M. Kanzler, J. Barth, J. Klucken, and B. M. Eskofier, "Inertial sensor based gait analysis discriminates subjects with and without visual impairment caused by simulated macular degeneration," Proc IEEE Eng Med Biol Soc, Orlando, FL, USA, Aug 16-20, 2016, pp. 4979-4982.

[6] S. Tedesco, A. Urru, J. Peckitt, and B. O'Flynn, "Inertial sensors-based lower-limb rehabilitation assessment: A comprehensive evaluation of gait, kinematic and statistical metrics," Int J Advances Life Sciences, 9, 1-2, pp. 33-49, 2017.

[7] S. Tedesco, A. Urru, and B. O'Flynn, "Spectral and time-frequency domains features for quantitative lower-limb rehabilitation monitoring via wearable inertial sensors," Proc IEEE Bio Circuits Systems, Turin, Italy, Oct 19-21, 2017.

[8] S. Tedesco, A. Urru, J. Peckitt, and B. O'Flynn 'Wearable Inertial Sensors as a Tool for Quantitative Assessment of Progress during Rehabilitation', Global Health 2016: The Fifth International Conference on Global Health Challenges, isbn: 978-1-61208-511-1

[9] S. Tedesco, A. Urru, A. Clifford, and B. O'Flynn, "Experimental validation of the Tyndall portable lower-limb analysis system with wearable inertial sensors," Procedia Engineering: 11th Conf Int Sport Eng Assoc, ISEA, The Engineering of Sport 11, 147, pp. 208-213, 2016.

[10] R. Houmanfar, M. Karg, and D. Kulic, "Movement analysis of rehabilitation exercises: distance metrics for measuring patient progress," IEEE Systems Journal, 10, 3, pp. 1014-1025, 2016.

[11] S. Fong, J. Liang, R. Wong, and M. Ghanavati, "A novel feature selection by clustering coefficients of variations," Int Conf Dig Inform Management (ICDIM), Phitsanulok, TH, Oct 1, 2014, pp. 205-213.

[12] T. Seel, J. Raisch, and T. Schauer, "IMU-based joint angle measurement for gait analysis," Sensors, 14, 4, pp. 6891-6909, 2016.

An Ontological Approach to Integrate Health Resources from Different Categories of Services

Cong Peng, Prashant Goswami, Guohua Bai

Department of Creative Technologies
Blekinge Institute of Technology
Karlskrona, Sweden
Email: cong.peng@bth.se, prashant.goswami@bth.se, guohua.bai@bth.se

Abstract—**Effective and convenient self-management of health requires collaborative utilization of health data from different services provided by healthcare providers, consumer-facing products and even open data on the Web. Although health data interoperability standards include Fast Healthcare Interoperability Resources (FHIR) have been developed and promoted, it is impossible for all the different categories of services to adopt in the near future. The objective of this study aims to apply Semantic Web technologies to integrate the health data from heterogeneously built services. We present an Web Ontology Language (OWL)-based ontology that models together health data from FHIR standard implemented services, normal Web services and Linked Data. It works on the resource integration layer of the presented layered integration architecture. An example use case that demonstrates how this method integrates the health data into a linked semantic health resource graph with the proposed ontology is presented.**

Keywords–Health data integration; ontology; FHIR; Semantic Web; Web service; eHealth; REST.

I. INTRODUCTION

Chronic diseases have become one of the main threats to people's health [1]. The caring of chronic diseases requires long-term and periodical management by both the patients and healthcare staff. The high cost and inconvenience of the chronic diseases caring make it better for the patients to perform self-management [2].

The development of information and communication technologies makes it much more feasible for health self-management. Electronic Health Record (EHR) systems have been adopted by many healthcare providers. Portable medical devices are used by patients for self-monitoring physiological parameters [2][3]. Moreover, many people today use wearable devices and health applications to record and manage their health [4][5].

The aforementioned systems, devices and applications record a huge amount of health data about patients. The collaborative utilization of the various health data has the potential to support chronic disease patients in having more effective and convenient self-management [6][7]. Unfortunately, all these health data became data silos, which can only be utilized in their own places with very limited outside collaboration. One reason behind this situation is that the systems holding these health data are heterogeneously built.

Web services technologies have promoted the interoperability of various software applications running on distributed and diversified systems. However, the lack of common standards adoption still makes it problematic for integrating health data with heterogeneous data models.

Various works have been done on health informatics standards, among which the Fast Healthcare Interoperability Resources (FHIR) created by the Health Level Seven International (HL7) organization is regarded as the next generation of health information interoperability framework that combines the previous standards' features [8][9]. It leverages common Web standards, which includes applying REpresentational State Transfer (REST) architectural style and JSON serialization format, besides the previous supported XML, as interfaces for health information exchange [10].

The FHIR working group is also working on the FHIR Linked Data module. It utilizes Semantic Web technologies include Resource Description Framework (RDF) and Web Ontology Language (OWL) to enhance its semantic expression capability and to facilitate inference and data linkage across datasets [8][10].

The work in this paper follows the path of FHIR to apply Semantic Web technologies and the REST resource model to integrate health data from different services as linked resources. We aim to link health services that adopted HL7 FHIR, implemented RESTful Web APIs or published as Linked Data, i.e., services with different levels of interoperability. The proposed approach is built on top of a semantic data aggregation method in [11]. The heterogeneous health data are modeled as conceptual information resources by using the Linked Health Resources (LHR) ontology. It makes the entire method a framework that aggregates health data from different sources and integrates them as health resources with semantics for upper level utilization.

The reminder of this paper is organized as follow. Section II introduces the previous works on integrating health data. Section III presents the method, which includes the health resource modeling, LHR ontology and the integration architecture. An example use case will be presented in Section IV to demonstrate the health data integration process by the proposed method. Finally, the paper will end with the conclusion section.

II. RELATED WORK

Many works have been done on integrating health data to support healthcare monitoring and decision making for either healthcare professionals or patients. In order to enable healthcare providers to remotely monitor and interpret health trends of diabetes patients, a method to integrate blood glucose data recorded from a patient-facing device to an EHR system was presented in [2]. The integration was achieved by transmitting the glucose data to the device vendor's iOS mobile phone application. It shares the data with the Apple HealthKit, which then sends the data to the EHR system. This solution is locked to the iOS platform since it depends on the Apple HealthKit.

V. Gay and P. Leijdekkers demonstrated a mobile application approach to aggregate health and fitness data to enable interoperability [5]. It was achieved by an Android application they developed with third-party partners to connect with wearable devices, EHR systems and other applications. A patient-centric mobile healthcare system that integrates data from body sensors was presented in [12]. The integration was implemented by leveraging a RESTful Web service on the application layer of the system to enable data sharing among applications.

There are also works done with Semantic Web technologies to enable health data integration with semantics. SENHANCE is a framework proposed by I. Pagkalos and L. Petrou to integrate patient self-reported health data on social network with hardware sensor observation data supported by Semantic Web technologies [3]. It models the former type of data as human sensor observations together with hardware sensor observations described by the ontology they proposed. B. Tilahun et al. presented a Linked Data-based system to retrieve and visualize heterogeneous health data in a flexible and reusable way [13]. The system utilized a set of Semantic Web technologies include RDF, Fuseki and SPARQL (SPARQL Protocol and RDF Query Language) for data representation, storage and query.

To integrate functionalities of different devices for supporting home-based care, an integration platform architecture was presented by Y. Trinugroho, F. Reichert and R. Fensli [14]. A smart home ontology that covered the modeling of person, device and context was proposed to enhance the reasoning process. To give semantics for the exchanged data between devices and systems in the home context, another smart home ontology was proposed in [15]. In addition, for the purpose of increasing the usability, the integration of smart home data with other data sources was also explored by applying Linked Data principles.

Despite there have been many works on using Semantic Web technologies to integrate health data, the integration between the ordinary health services and the services implemented with the next generation standard FHIR remains unexplored.

III. METHODS

The LHR framework intends to integrate 3 categories of health data sources: FHIR-enabled services, health services implemented Web APIs (satisfy certain REST constraints) and health related Linked Data. The 3 categories cover most available health related services and data, either for the current stage or the near future development trends.

A. The Nature of Health Resource

Today most of the health services that provide data access are providing Web APIs, which usually serve data in JSON or XML serialization format via HTTP methods. Most of these Web APIs claim to be RESTful services. Though in many cases they only follow a few of the REST constraints, which makes them actually not RESTful services. However, most of the services follow one of the fundamental elements of REST to organize the accessible data as resources, which is the unit of information in REST architectural style [16]. It makes them capable of being modeled together under the LHR framework.

In order to improve the implementability and to be more developer-friendly, the HL7 FHIR standard leverages the common Web technologies and concepts. FHIR therefore follows the REST architectural style as well, and is built upon a set of resources. Resource in the case of FHIR means "a collection of information models that defines the data elements, constraints and relationships for the objects relevant to healthcare" [8]. The objects that are modeled as resources include *Patient*, *Observation*, *OperationOutcome* and so on. Each resource is defined in a certain structure with references to other resources, and represented in XML, JSON and an additional RDF serializable format Turtle.

Linked Data is a practice of publishing structured and interlinked data with semantic meanings on the Web to make it a Web of data [17]. It was proposed by Tim Berners-Lee as an application of Semantic Web technologies. The Linked Data rules align well with some of the REST constraints. There have been many works on linking REST Web services and Linked Data [18][19]. Different levels of practices exist on publishing Linked Data according to the 5 star rating system developed by Berners-Lee [17]. A proper practice of publishing Linked Data should identify interlinked things (data items or real world entities represented on the Web) with HTTP Uniform Resource Identifiers (URIs) and serve corresponding information in RDF serializable formats or SPARQL query service.

Accordingly, we can regard anything identified by an HTTP URI as a resource, i.e., a node in an RDF graph. And the resource identified by the root path of an HTTP URI can be regarded as a service, i.e., a root node of an RDF graph.

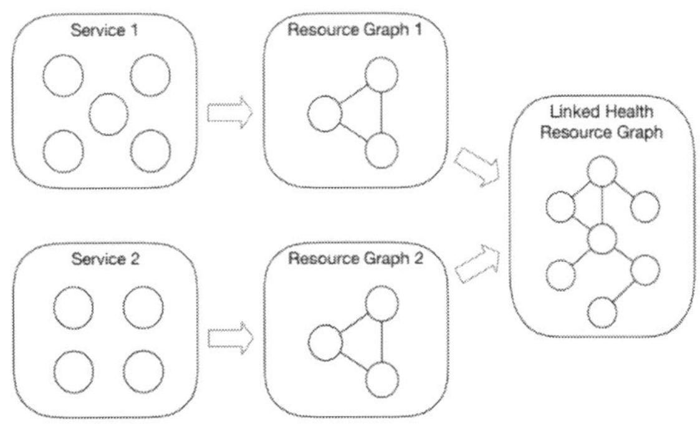

Figure 1. A simple overview of the integration process (a circle stands for a resource), resources are aggregated from services into semantic resource graphs, which are then integrated as a Linked Health Resource graph

B. Modeling of Health Resource

Based on the nature of health Web services and Linked Data services, each service contains a set of resources, no the service is implemented with FHIR or not. Let S_i be a health service, where $S_i = \sum R_j$ denotes that service S_i serves a set of resources R_j.

We assume that a person P_k has several health services, which have been integrated together into his or her health

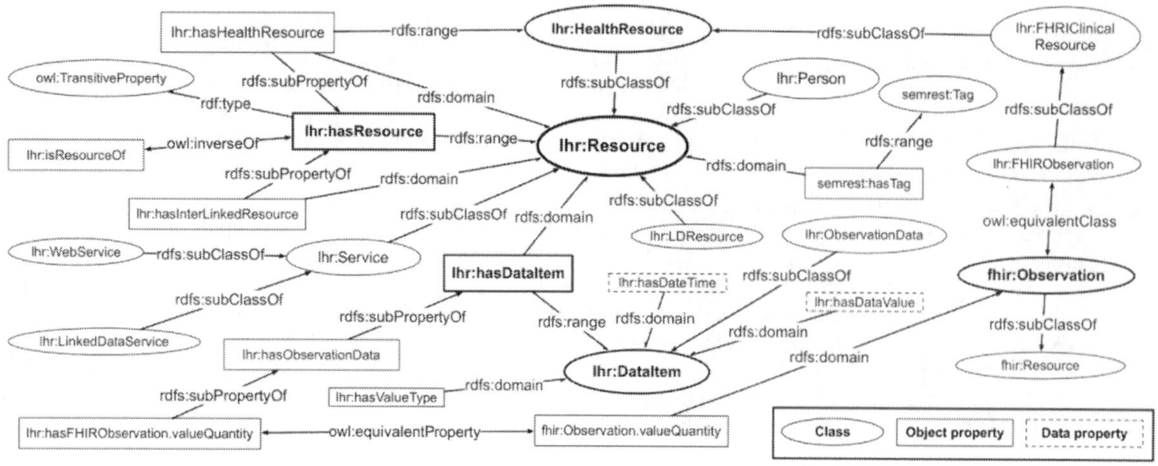

Figure 2. Linked Health Resource Ontology (with vital objects), the integrated health services and resources are modeled as sub-classes of the *lhr:Resource* class, which is the fundamental class in this ontology. Certain equivalent classes and properties are mapped to FHIR vocabulary for the integration of FHIR resources

resource graph $G(P_k)$. A health resource graph means a set of health information being modeled as resources with relationships, which maps to the RDF graph data model. Then $G(P_k) = \sum S_i = \sum\sum R_{ij}$ denotes that a person's health status can be described by sets of health resources from a collection of services. The sets of health resources can be represented in an integrated health resource graph. Figure 1 shows a simple overview of the process that resources from different services being integrated into a health resource graph.

Furthermore, a resource itself contains several objects:

- A resource may have interlinked (subordinate or related) resources R_{link}.
- A resource may have data items D_r that describe certain status about a person, for example profile data, observation data, etc.
- A resource can be annotated by a set of n semantic tags T_r, where $n \in [0,\infty]$.

A resource can therefore be represented as a tuple $R_j = (D_r, R_{link}, T_r)$.

C. Linked Health Resource Ontology

When we look into the very nature of the relationships among person, service and resource, we can see that a service as a unit also has resources that are interlinked, which is the same as a resource. It makes a service to be a resource. The same applies to a person, it then makes a person to be a resource as well. Therefore, we model the three types of objects all as sub-classes of class *lhr:Resource* under the LHR ontology, where the basic relationships among them being *lhr:hasResource* and *lhr:isResourceOf*. Figure 2 shows the main classes and properties in the LHR ontology. It is built via the W3C recommendation standards, the RDF Schema (RDFS) data modeling vocabulary [20] and the OWL ontology language [21].

As we have stated, the class *lhr:Resource* is the super-class of all the involved resources. The class *lhr:Service* is a sub-class of *lhr:Resource*, and represents a set of resources in a relatively separated territory to indicate a source of health data. The class *lhr:Person* represents a person who owns all the health resources. The class *lhr:HealthResource* represents the person instance's all health resources that are interlinked from normal health Web services and FHIR implemented services. Usually Linked Data are not about only a single person, so resources from Linked Data is classified as the more generic class *lhr:LDResource*.

To link the FHIR health-related resources, which are categorised as clinical resources in FHIR, we map those clinical resources into the LHR ontology. For instance, the *fhir:Observation* is mapped to *lhr:FHIRObservation*, which is a sub-class of *lhr:HealthResource*. Health resources from other normal Web services or Linked Data sets can also map from a class that have been defined in other ontologies or vocabularies to a sub-class of *lhr:HealthResource*. Alternatively, the origin resource can instantiate directly from *lhr:HealthResource* if there is no embedded class.

Object properties around the class *lhr:Resource* and its sub-classes are the transitive property *lhr:hasResource* and its sub-properties include *lhr:hasHealthResource*, *lhr:hasLDResource* and *lhr:hasInterLinkedResource*. Among which, the property *lhr:hasHealthResource* has *lhr:Resource* as its domain and *lhr:HealthResource* as its range, the property *lhr:hasLDResource* has *lhr:Resource* as its domain and *lhr:LDResource* as its range.

The class *lhr:DataItem* is used to represent the data model of a *lhr:Resource* if it has no predefined model, and class *lhr:ObservationData* is used to represent the health data model of a *lhr:HealthResource*. Resources from FHIR service have predefined data models in RDF, therefore, the object of which are mapped to an instance of *lhr:ObservationData*. The relation is represented by a sub-property of the object property *lhr:hasObservationData*.

With the sub-class relations and transitive sub-property relations defined in the LHR ontology, we can easily infer an instance of *lhr:Person* has a set of instances of *lhr:HealthResource* being linked together without complex inference rules.

D. Layered Architecture towards Integration

For integrating different health data into one ontological model with LHR, we need to firstly aggregate the health data together from different services. Figure 3 illustrates the layered architecture for integrating health data from different categories of services, as aforementioned that the LHR framework intends to integrate, into a LHR ontological model.

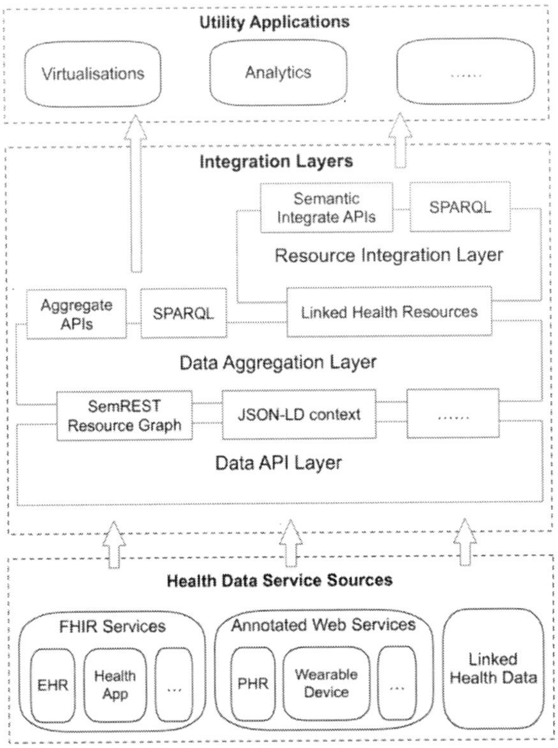

Figure 3. Layered architecture of health data integration with Linked Health Resource. The ontology modeling process works on the resouce integration layer, which receives the aggregated resource graph from the data aggregation layer

From bottom to top, the first layer is the data API layer. This layer simply requests data from services via APIs, e.g., Web APIs or Linked Data APIs. The retrieved data are then aggregated together into a semantic resource graph on the data aggregation layer.

In order to be effectively aggregated into the semantic resource graph, it is necessary for the normal Web services to have some simple semantic annotations with commonly used vocabularies. Besides the Semantic RESource Tagging (SemREST) in [22], JSON-LD context embedding to ordinary JSON representation [23] or other Semantic Web service annotation methods such as the ones presented in [24]–[26] could also be used to semantically annotate the resource.

As FHIR implemented services follow the FHIR standards, therefore, the data resources from these services come with standard structure and semantics. Some of the services even implemented FHIR/RDF representations. Data retrieved from Linked Data services are already RDF serializable if they were implemented properly. So the two categories of services can be aggregated into a semantic resource graph without much effort.

The semantic resource graph aggregated with the retrieved health data is then sent to the information resource integration layer. The LHR ontology is used to extract health resources from the resource graph for integration. One thing that needs to be noted is that the concept *resource* in the semantic resource graph from the data aggregation layer is slightly different from the *lhr:Resource* in LHR ontology. The former *resource* maps to a resource in a REST Web service, it contains functional meta information of its Web API. Only its representation unit will be integrated into the integration layer as an instance of *lhr:HealthResource*.

On top of the information resource integration layer there could be semantic APIs that utilize the health resource instances of *lhr:Resource* and the contained *lhr:DataItem*. Providing APIs on modelized resources in this manner has the potential to provide a unified interface to utilize the health data from different sources in the wrapped model for applications like health data visualization, analytics and so on. Alternatively, applications could also utilize health data in the form of semantic resource graph above the data aggregation layer. Thanks to all the resources are represented in RDF on both the aggregation layer and integration layer, it is therefore able to serve SPARQL queries directly.

IV. A Use case of Integration

To demonstrate the LHR health data integration process, an example use case will be presented in this section. It is depicted as an idealized scenario where it leaves aside some details like the authentication and invocation of services. The demonstration prototype was implemented as two parts: the *aggregation layer* was implemented in Python with RDFlib package, and the *integration layer* was implemented in Java with Jena framework. The example use case that will be presented focuses on the integration layer since it is the main content of this paper. Therefore, the working process of aggregation is omitted here.

In a scenario that a patient named Alice has taken some medical examination at a healthcare center, which provides health data access via an HL7 FHIR implemented service. At the same time, she also has been using a Fitbit health band to track her daily activity and life style data for some time. In order to have a periodical comprehensive view of her health, she wants to integrate her *blood glucose* examination data from the healthcare center and her data of *daily steps performed* together.

For demonstration purpose, the data sample of *blood glucose* uses the FHIR official observation example [27]. Listing 1 shows the simplified resource of Alice's *blood glucose* observation data in RDF/Turtle format. This resource is passed through the data aggregation layer to the integration layer directly since the FHIR observation resource is an instance of the class *fhir:Observation* that maps to *lhr:FHIRObservation*.

Next, we are going to integrate the *daily steps performed* data. Fitbit provides Web APIs to access the data recorded by their wearable devices. For integrating the data from Fitbit Web service into the LHR model, we choose to apply the SemREST method [11][22] to annotate its service description and aggregate the resource representation on the data aggregation layer. The generated semantic resource graph of Fitbit *daily steps performed* is partly shown in Listing 2 in RDF/Turtle format. It is then passed to the integration layer.

```
@prefix fhir: <http://hl7.org/fhir/> .
@prefix loinc: <http://loinc.org/owl#> .
@prefix xsd: <http://www.w3.org/2001/XMLSchema#> .

<http://hl7.org/fhir/Patient/f001> a fhir:Patient .

<http://hl7.org/fhir/Observation/f001> a fhir:Observation;
  fhir:Observation.identifier [
     fhir:Identifier.value [ fhir:value "6323" ]
  ];
  fhir:Observation.subject [
     fhir:link <http://hl7.org/fhir/Patient/f001>;
  ];
  fhir:Observation.code [
     fhir:CodeableConcept.coding [ a loinc:15074-8; ]
  ];
  fhir:Observation.effectiveDateTime [
     fhir:value "2018-01-14T09:30:10+01:00"^^xsd:dateTime
  ];
  fhir:Observation.valueQuantity [
     fhir:Quantity.value [ fhir:value "6.3"^^xsd:decimal ];
     fhir:Quantity.unit [ fhir:value "mmol/l" ];
  ];
  fhir:Observation.interpretation [
     fhir:CodeableConcept.coding [
       fhir:Coding.code [ fhir:value "H" ];
       fhir:Coding.display [ fhir:value "High" ]]
  ] .
```

Listing 1. Simplified FHIR observation of blood glucose example represented in RDF/Turtle format. The observation resource contains the vlaue, the effective date time and the LOINC code indicating the medical terminology

```
@prefix fitbit: <http://fitbit.com/namespace#> .
@prefix schema: <http://schema.org/> .
@prefix semrest: <http://semrest.org/vocab#> .
@prefix xsd: <http://www.w3.org/2001/XMLSchema#> .

<http://api.fitbit.com/> a semrest:Description ;
    semrest:hasResource fitbit:resource1.

fitbit:resource1 a semrest:Resource ;
    semrest:hasResponse fitbit:response1 ;
    semrest:hasMethod semrest:get ;
    semrest:hasTag schema:steps,
        schema:activities;
    semrest:urlTemplate "/user/activities/steps/date/
 ↪  {base-date}/{end-date}.json"
 ↪  .

fitbit:response1 a semrest:Response;
    semrest:hasRepresentation fitbit:representation1 .

fitbit:representation1 a semrest:Representation ;
    semrest:dataItem [ semrest:hasValue "7430" ;
            semrest:hasTag schema:steps ;
            schema:dateTime "2018-01-12"^^xsd:date ],
        [ semrest:hasValue "3324" ;
            semrest:hasTag schema:steps ;
            schema:dateTime "2018-01-13"^^xsd:date ],
        [ semrest:hasValue "4739" ;
            semrest:hasTag schema:steps ;
            schema:dateTime "2018-01-14"^^xsd:date ],
        [ semrest:hasValue "9496" ;
            semrest:hasTag schema:steps ;
            schema:dateTime "2018-01-15"^^xsd:date ].
```

Listing 2. Simplified resource graph of Fitbit steps resource generated by SemREST and represented in RDF/Turtle format. The resouce contains each day's steps count, date time and the steps term in schema.org as a tag indicating the semantic meaning

```
PREFIX rdf: <http://www.w3.org/1999/02/22-rdf-syntax-ns#>
PREFIX lhr: <http://lhr.org/ontology/2018/4/ns#>

SELECT ?resource ?dataItem ?value ?dateTime ?valueType

WHERE {
  <:Alice> lhr:hasResource ?resource .
  ?resource rdf:type lhr:HealthResource .
  ?resource lhr:hasDataItem ?dataItem .
  ?dataItem lhr:hasDataValue ?value .
  ?dataItem lhr:hasDateTime ?dateTime .
  ?dataItem lhr:hasValueType ?valueType.
}
```

Listing 3. Example SPARQL query to retrieve integrated LHR health resources with their data value, date time and value type

In the integration layer, the coming resources are integrated using the LHR ontology. As mentioned before, the FHIR *blood glucose* observation resource is an instance of *lhr:FHIRObservation*, which is an equivalent class to *fhir:Observation*. For the ease of accessing data value of FHIR observation resources with LHR, the property *fhir:Observation.valueQuantity* is mapped to the corresponding object property *lhr:hasFHIRObservation.valueQuantity*, which is a sub-class of *lhr:hasObservationData*. And the *fhir:Observation.effectiveDateTime* is mapped to the corresponding data property *lhr:hasObservationDateTime*. Other objects could be accessed directly through the *fhir:Observation* instance, including the *fhir:Observation.code*, *fhir:Observation.interpretation* and *fhir:Observation.referenceRange*. Furthermore, for a more convenient utilization for upper level APIs, the object of *fhir:Observation.code* could be mapped to *lhr:DataItem*'s value type to indicate the semantic meaning of the data value.

For the Fitbit *daily steps performed* resource, the *semrest:Representation* instance *fitbit:representation1* in the resource graph from the aggregation layer is instantiated as a *lhr:HealthResource*. Since the *steps* data has been annotated as the property *semrest:dataItem*'s objects, so it can be mapped to the objects of *lhr:hasObservation*, a sub-property of *lhr:hasDataItem*, to this LHR health resource. The data value, date time and tag could also be mapped to the objects in the ontology for the upper level APIs.

The RDF of the integrated health resources is depicted graphically in Figure 4 with relatively important objects. Each square block represents an instance with its class and super-class. Instances are linked by object properties and data properties.

As one of the methods that can utilize the integrated health resources, a simple SPARQL query in Listing 3 is executed to query for Alice's integrate health resources. The query result is presented in Figure 5. We can see that both the FHIR observation resource and the Fitbit resource are retrieved, since both the two types of resources were modeled as *lhr:HealthResource* directly or through sub-class relation. As the value, date time and value type have been mapped into the LHR ontology, so we can query them with the properties of *lhr:DataItem*. The value type of each data item is actually an URI. It not only works as the value type's dereferencable URI for indicating the semantic meaning, but also integrates the resource under LHR with other Linked Data, where the

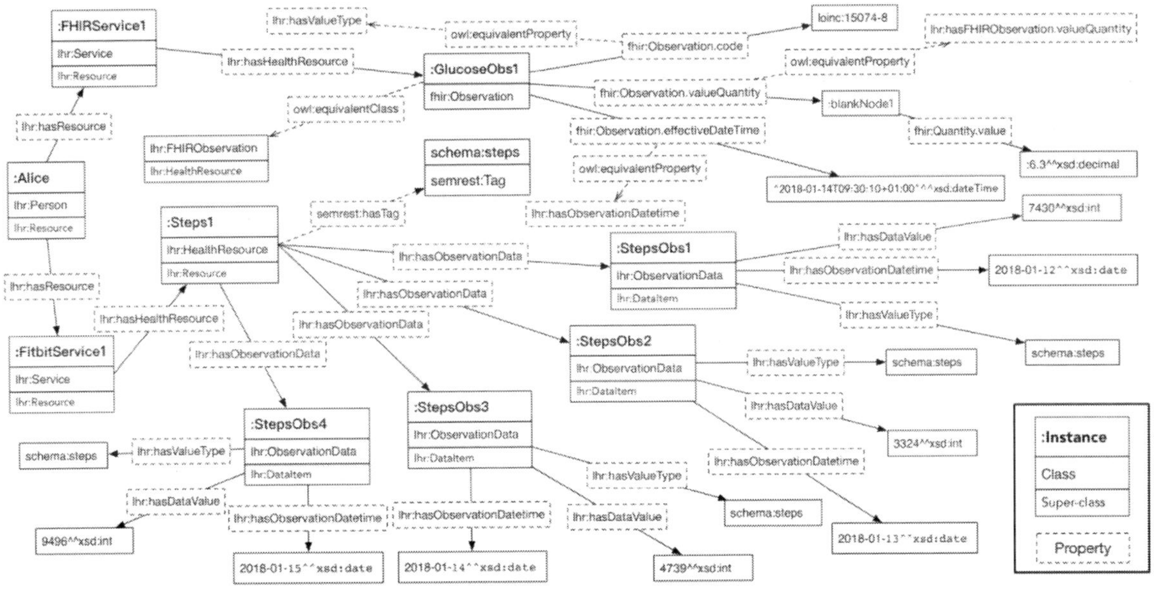

Figure 4. Graphical presentation of Alice's integrated health resources under the LHR ontology, where each instance's class or property is presented with super-class or super-property.

	resource	dataItem	value	dateTime	valueType
1	http://fitbit.com/namespace#representation1	b2	7430	"2018-01-12"^^xsd:date	http://schema.org/steps
2	http://fitbit.com/namespace#representation1	b3	3324	"2018-01-13"^^xsd:date	http://schema.org/steps
3	http://fitbit.com/namespace#representation1	b0	4739	"2018-01-14"^^xsd:date	http://schema.org/steps
4	http://hl7.org/fhir/Observation/f001	b4	"6.3"^^xsd:decimal	"2018-01-14T09:30:10+01:00"^^xsd:dateTime	http://loinc.org/owl#15074-8
5	http://fitbit.com/namespace#representation1	b1	9496	"2018-01-15"^^xsd:date	http://schema.org/steps

Figure 5. The result of the example SPARQL query, all the Alice's resources that were modeled as *lhr:HealtResource* are retrieved with their data value, data time and value type

same URI were referenced.

In order to demonstrate the health data integration in a simple view, the example use case was performed far from a realistic way of managing health. However, it is capable to work in real cases since the data integration process is the same.

V. Conclusion

In this paper, we presented a method to integrate HL7 FHIR interoperability standard implemented health services and normal Web based health services with a few semantic annotations. The proposed OWL based LHR ontology modeled different categories of health resources together in a clear simple way for upper level unified utilization. Thanks to the application of Semantic Web technologies and the well aligned conceptual model with Linked Data, this method can also integrate health resources from the Linked Open Data on the Web without much effort.

With the capability to integrate different categories of health services, which usually contain health data from health-care providers, consumer-facing products and health research, the presented method has the potential to support self-management of health especially for people with chronic diseases. To achieve a satisfied interoperable integration, it is ideal that the involved health services could implement interoperability standards such as HL7 FHIR. However, obstacles include services with legacy system problems make it impossible for all the services to realize in the near future. Diversity may exist among the different health services. The approaches like the one we proposed can act as a bridge to connect the heterogeneously implemented services. Integrating health data as an ontological model with semantic meaning is beneficial for upper level utilization in a more unified way.

Limitations, however, exist in this work. It still requires extra semantic annotation by the service providers for integrating the normal Web health services. Otherwise, it is unable to identify which part of the resource should be integrated. The more semantic annotations embedded in a resource's representation, the more machine-understandable integration it will be, which can promote more useful applications. Manual ontology mapping is also needed for the resources from the data aggregation layer to be integrated on the integration layer.

For future works, we plan to evaluate this method with

more real world services for improvement and validation. In addition, the upper level semantic utilization APIs and applications will be explored for the collaborative utilization of health data.

REFERENCES

[1] A. S. Daar, P. A. Singer, D. Leah Persad, S. K. Pramming, D. R. Matthews, R. Beaglehole, A. Bernstein, L. K. Borysiewicz, S. Colagiuri, N. Ganguly, R. I. Glass, D. T. Finegood, J. Koplan, E. G. Nabel, G. Sarna, N. Sarrafzadegan, R. Smith, D. Yach, and J. Bell, "Grand challenges in chronic non-communicable diseases," Nature, vol. 450, no. 7169, nov 2007, pp. 494–496. [Online]. Available: http://www.nature.com/articles/450494a

[2] R. B. Kumar, N. D. Goren, D. E. Stark, D. P. Wall, and C. A. Longhurst, "Automated integration of continuous glucose monitor data in the electronic health record using consumer technology," Journal of the American Medical Informatics Association, vol. 23, no. 3, may 2016, pp. 532–537. [Online]. Available: https://academic.oup.com/jamia/article-lookup/doi/10.1093/jamia/ocv206

[3] I. Pagkalos and L. Petrou, "SENHANCE: A Semantic Web framework for integrating social and hardware sensors in e-Health," Health Informatics Journal, vol. 22, no. 3, sep 2016, pp. 505–522.

[4] P. Vuorimaa, P. Harmo, M. Hämäläinen, T. Itälä, and R. Miettinen, "Active life home," in Proceedings of the 5th International Conference on PErvasive Technologies Related to Assistive Environments - PETRA '12. New York, New York, USA: ACM Press, 2012, p. 1. [Online]. Available: http://dl.acm.org/citation.cfm?doid=2413097.2413133

[5] V. Gay and P. Leijdekkers, "Bringing Health and Fitness Data Together for Connected Health Care: Mobile Apps as Enablers of Interoperability," Journal of medical Internet research, vol. 17, no. 11, jan 2015, p. e260.

[6] S. Rea, J. Pathak, G. Savova, T. A. Oniki, L. Westberg, C. E. Beebe, C. Tao, C. G. Parker, P. J. Haug, S. M. Huff, and C. G. Chute, "Building a robust, scalable and standards-driven infrastructure for secondary use of EHR data: The SHARPn project," Journal of Biomedical Informatics, vol. 45, no. 4, aug 2012, pp. 763–771. [Online]. Available: http://www.sciencedirect.com/science/article/pii/S1532046412000202

[7] N. Philip, T. Butt, D. Sobnath, R. Kayyali, S. Nabhani-Gebara, B. Pierscionek, I. Chouvarda, V. Kilintis, P. Natsiavas, N. Maglaveras, and A. Raptopoulos, "Design of a RESTful middleware to enable a web of medical things," in Proceedings of the 2014 4th International Conference on Wireless Mobile Communication and Healthcare - "Transforming Healthcare Through Innovations in Mobile and Wireless Technologies", MOBIHEALTH 2014, 2015, pp. 361–364.

[8] HL7, "Fast Healthcare Interoperability Resources (FHIR)," Accessed: 2018-02-07. [Online]. Available: https://hl7.org/fhir/

[9] R. A. Bloomfield, F. Polo-Wood, J. C. Mandel, and K. D. Mandl, "Opening the Duke electronic health record to apps: Implementing SMART on FHIR," International Journal of Medical Informatics, vol. 99, 2017, pp. 1–10.

[10] M. P. Luz, J. R. d. M. Nogueira, L. T. Cavalini, and T. W. Cook, "Providing Full Semantic Interoperability for the Fast Healthcare Interoperability Resources Schemas with Resource Description Framework," in 2015 International Conference on Healthcare Informatics. IEEE, oct 2015, pp. 463–466. [Online]. Available: http://ieeexplore.ieee.org/document/7349736/

[11] C. Peng, P. Goswami, and G. Bai, "Linking Health Web Services as Resource Graph by Semantic REST Resource Tagging," in The 8th International Conference on Current and Future Trends of Information and Communication Technologies in Healthcare (ICTH). Procedia Computer Science, 2018.

[12] J. Choi, M. Ha, J. Im, J. Byun, K. Kwon, W. Yoon, D. Kim, S. Heo, and D. Kim, "The patient-centric mobile healthcare system enhancing sensor connectivity and data interoperability," in 2015 International Conference on Recent Advances in Internet of Things, RIoT 2015. IEEE, apr 2015, pp. 1–6. [Online]. Available: http://ieeexplore.ieee.org/document/7104911/

[13] B. Tilahun, T. Kauppinen, C. Keßler, and F. Fritz, "Design and development of a linked open data-based health information representation and visualization system: potentials and preliminary evaluation," JMIR medical informatics, vol. 2, no. 2, jan 2014, p. e31. [Online]. Available: http://www.pubmedcentral.nih.gov/articlerender.fcgi?artid=4288106{&}tool=pmcentrez{&}rendertype=abstract

[14] Y. B. D. Trinugroho, F. Reichert, and R. Fensli, "An Ontology-Enhanced SOA-Based Home Integration Platform for the Well-Being of Inhabitants," in 4th IADIS International Conference on e-Health, 2012, pp. 159–164.

[15] J. Puustjärvi and L. Puustjärvi, "The Role of Smart Data in Smart Home: Health Monitoring Case," Procedia Computer Science, vol. 69, no. 7th International Conference on Advances in Information Technology, jan 2015, pp. 143–151. [Online]. Available: https://www.sciencedirect.com/science/article/pii/S1877050915031798

[16] R. T. Fielding, "Architectural styles and the design of network-based software architectures," Ph.D. dissertation, University of California, Irvine, 2000. [Online]. Available: http://dl.acm.org/citation.cfm?id=932295

[17] T. Berners-Lee, "Linked Data - Design Issues," Accessed: 2018-03-23, 2009. [Online]. Available: https://www.w3.org/DesignIssues/LinkedData.html

[18] R. Alarcon and E. Wilde, "Linking Data from RESTful Services," in Third Workshop on Linked Data on the Web, 2010. [Online]. Available: http://citeseerx.ist.psu.edu/viewdoc/summary?doi=10.1.1.183.1253

[19] M. Lanthaler and C. Gütl, "A semantic description language for RESTful Data Services to combat Semaphobia," in 5th IEEE International Conference on Digital Ecosystems and Technologies (IEEE DEST 2011), may 2011, pp. 47–53.

[20] "RDF Schema 1.1 | W3C," Accessed: 2018-03-02. [Online]. Available: https://www.w3.org/TR/rdf-schema

[21] "OWL 2 Web Ontology Language | W3C," Accessed: 2018-03-03. [Online]. Available: https://www.w3.org/TR/owl-syntax/

[22] C. Peng and G. Bai, "Using Tag based Semantic Annotation to Empower Client and REST Service Interaction," in Proceedings of 3rd International Conference on Complexity, Future Information Systems and Risk (COMPLEXIS2018), 2018, pp. 64–71.

[23] M. Lanthaler and C. Gütl, "On using JSON-LD to create evolvable RESTful services," in Proceedings of the Third International Workshop on RESTful Design - WS-REST '12, 2012, p. 25. [Online]. Available: http://dl.acm.org/citation.cfm?doid=2307819.2307827

[24] M. N. Lucky, M. Cremaschi, B. Lodigiani, A. Menolascina, and F. De Paoli, "Enriching API Descriptions by Adding API Profiles Through Semantic Annotation," in International Conference on Service-Oriented Computing. Springer, Cham, 2016, pp. 780–794. [Online]. Available: http://link.springer.com/10.1007/978-3-319-46295-0{_}55

[25] L. Panziera and F. De Paoli, "A framework for self-descriptive RESTful services," in Proceedings of the 22nd International Conference on World Wide Web - WWW '13 Companion, 2013, pp. 1407–1414. [Online]. Available: http://dl.acm.org/citation.cfm?doid=2487788.2488183

[26] R. Verborgh, T. Steiner, D. Van Deursen, R. Van De Walle, and J. G. Valles, "Efficient runtime service discovery and consumption with hyperlinked RESTdesc," in Proceedings of the 2011 7th International Conference on Next Generation Web Services Practices, NWeSP 2011, 2011, pp. 373–379.

[27] HL7, "Resource Observation - Examples," Accessed: 2018-02-16. [Online]. Available: https://www.hl7.org/fhir/observation-examples.html

dokspot – Securely Linking Healthcare Products with Online Instructions

Kevin Lapagna, Moritz Zollinger, Marc Rennhard
School of Engineering
Zurich University of Applied Sciences
Winterthur, Switzerland
Email: lkev,zolg,rema@zhaw.ch

Hans Strobel, Cyrille Derché
dokspot GmbH
Zurich, Switzerland
Email: hans.strobel,cyrille.derche@dokspot.com

Abstract—**Printed instructions for products get replaced more and more by digital versions that are made available over the Internet. In safety-sensitive fields, such as healthcare products, availability and integrity of these instructions is of highest importance. However, providing and managing instructions online opens the door to a wide range of potential attacks, which may negatively affect availability and integrity. In this paper, dokspot is presented, which is an Internet-based service that aims at solving this problem by securely linking healthcare products with online instructions. The key to achieve this is a sophisticated security architecture and the focus of this paper is on the core components of this architecture. This includes a secure workflow to manage online instructions, which prevents, e.g., attacks by malicious insiders. Also, the traditionally monolithic web application architecture was split into role-based microservices, which provides protection even if parts of the system are compromised. Furthermore, digital signatures are utilized to continuously safeguard the lifecycle of online instructions to guarantee their genuineness and integrity. And finally, a passwordless signature scheme is introduced to hide inconvenient extra steps from the users while still maintaining security. Overall, this security architecture makes dokspot highly resistant to a wide range of attacks.**

Keywords–Web Application Security; Microservices; Digital Signatures; Passwordless Signatures; Healthcare Product Instructions; Online Document Management System.

I. Introduction

Today, products should be designed in a way that allows intuitive and safe use without reading the instructions. However, with increasing risks of using a product, relying on intuition and using a "trial and error" approach is unacceptable and becomes a potential safety risk. An illustrative example is the operation of a passenger airplane, where pilots have to read and tick off instructions every time before operating the aircraft. Everybody would agree that in this scenario, doing it in this way and by using the right instructions is an important safety factor.

Another product category that requires detailed knowledge of the instructions are healthcare products, which – in the context of this paper – includes any substance, product, or system used for therapeutic or diagnostic purposes on the human body (or animals). Healthcare professionals (doctors, nursing staff, operators of medical machines, etc.) must be aware of all details involving the use of a healthcare product prior to its application on a patient. To achieve this, the healthcare professional must have guaranteed and simple access to the right instructions in the right language at the right time.

To comply with this, manufacturers of healthcare products predominantly ship printed instructions in multiple languages together with their products. This has various disadvantages, including that the related costs – financial, operational, and environmental – are substantial, that most of the included instruction languages remain unused, and that the required instructions can often not be found when needed (e.g., because they were misplaced or thrown away). Also, it is sometimes necessary to break a product seal to get to the printed instructions, which implies the product can often not be returned to the manufacturer if it turns out the product is not suited for the planned application.

Some of these limitations can be remedied by providing the instructions online. This is typically done with a web portal where healthcare professionals can search for and download instructions of specific products. While this sounds to be a good solution, it has its limitations in practice. One limitation is that every manufacturer uses its own portal, which means the healthcare professional not only has to find the right portal, but also has to be able to cope with different user interfaces. Another limitation is that once the right portal has been found, it may be difficult to find the correct instruction for a specific product. A third limitation is that such a portal may be an attractive attack point, e.g., for competitors or for outside attackers, as tampering with the provided instructions in malicious ways may have a devastating effect on patient safety.

To overcome these limitations, dokspot was developed. dokspot is a novel Internet-based service that aims at transforming the way companies handle instructions by providing an innovative paperless and trustworthy solution. Figure 1 depicts the basic functionality of dokspot.

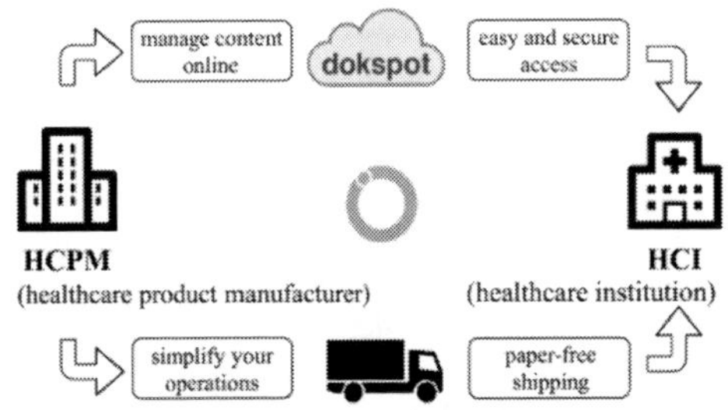

Figure 1. Basic Functionality of dokspot

Figure 1 shows that manufacturers of healthcare products

(left side, these are the customers of the dokspot service) upload their instructions to dokspot (top). Users of the products, in this example healthcare professionals that are typically working at healthcare institutions (right side, these are the customers of the manufacturer of the healthcare products), can easily access this information using the dokspot service when they need to do so. Of course, the physical products must still be shipped to the healthcare institutions (bottom), but they do not include instructions on paper. Among other advantages, this simplifies operations for the manufacturers of healthcare products and reduces hospital waste.

To link a physical product to the online instructions, a Uniform Resource Locator (URL) pointing to the dokspot service is used. This URL is defined by the manufacturer of the healthcare product and communicated to the healthcare professional (the user of the device) together with the product, e.g., on the product label. Figure 2 shows an example of such a label, which includes the URL on the right side. The important contents of this label are a product-specific alphanumeric code and a scannable data matrix.

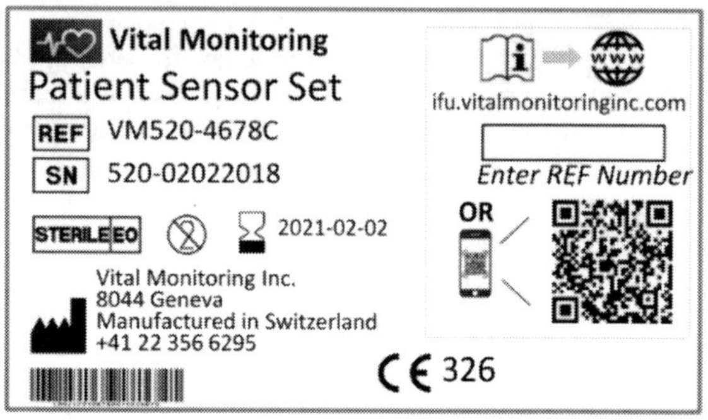

Figure 2. Example of a dokspot Label.

To access the instructions, the healthcare professional simply scans the data matrix (using for example a standard QR code reader app on a smartphone or tablet) or enters the product-specific code (here: VM520-4678C) on the web page at the URL provided by the manufacturer (here: ifu.vitalmonitoringinc.com). As a result, the instructions are displayed in the required language.

It is obvious that this approach removes some of the limitations identified earlier. However, there is still the challenging problem of how such a service should be designed and developed so that it provides sufficient protection from attacks. In particular, it must be guaranteed that when a user accesses online instructions through dokspot, there is high assurance that the instructions they receive and view on their device match the original instructions as provided by the manufacturer of the healthcare product. In this context, it is also important to realize that several regulatory authorities have issued guidelines obliging manufacturers to protect the integrity of their electronic files, which includes protecting online instructions from unauthorized modification. One recent example for such a guideline is the GXP Data Integrity Definitions and Guidance from the UK Medicine & Healthcare Products Regulatory Agency (MHRA) [1].

To our knowledge, no work has been published so far that presents a solution to this challenging problem. The focus of this paper therefore lies on filling this gap by presenting the underlying security architecture of dokspot. This security architecture contains several key components to protect from a wide range of attacks and provides the basis to achieve the required level of security and also to comply with regulations as mentioned earlier. As a result of this, dokspot truly allows to securely link healthcare products with online instructions. Note that while this security architecture has been designed, developed and evaluated in the context of dokspot and for healthcare products, the architecture is general enough so it can be applied to other industries as well. Therefore, the main contribution of this paper is to provide a security architecture that allows to securely link physical products and online documents in general, independent of the field of application.

The remainder of this paper is organized as follows: Section II describes the security goals and threat model of dokspot. Based on this, the security architecture is explained in Section III, followed by the evaluation in Section IV. Related work is covered in Section V and Section VI concludes this work.

II. Security Goals and Threat Model

In this section, the overall architecture, the security goals and the threats against dokspot are described. This provides the basis for the security architecture that follows in Section III.

A. *Overall Architecture*

Figure 3 depicts a high-level view of the overall architecture of dokspot.

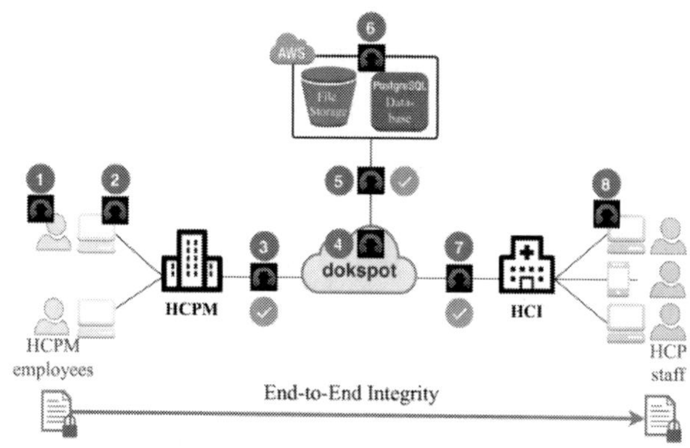

Figure 3. dokspot Architecture and Threats against dokspot.

In the middle of Figure 3, there is the central component of the dokspot service. This component corresponds to a web application that provides the core functionality and which is deployed on a suitable cloud application platform. To store the instructions, a cloud storage provider is used. Using 3rd party services to run the application is much more reasonable than deploying and operating an own infrastructure, especially as these services provide high availability and can easily scale with growing demands with respect to computing power and storage space.

On the left side, there is a company – identified as healthcare product manufacturer (HCPM) – that uses dokspot to provide its instructions. HCPM employees are interacting with the dokspot service via the web browser on their workplace computers to upload and maintain instructions of their products. To support this, the dokspot service provides a corresponding web interface.

On the right side, there is a healthcare institution (HCI) that is using the online instructions provided by the HCPMs via dokspot. Its users are identified as healthcare professionals (HCP). HCPs can access the instructions using any web browser on any device, including workstations, laptops, tablets or smartphones. This is also supported with a corresponding web interface provided by the dokspot service.

B. Security Goals

With respect to the general security goals – confidentiality, integrity, and availability – dokspot is a "special case". In many typical Internet-based applications, confidentiality is highly important, but in the case of dokspot, confidentiality of the instructions is not critical at all. All instructions provided through dokspot are usually considered public information and it is therefore not necessary to restrict read-access to them.

Availability and integrity are paramount, though. The instructions should be available within a few seconds whenever they are needed by a HCP and unavailability may mean – in the worst case – that patient safety is negatively affected. High availability is achieved by operating the service using well-established cloud-based services that have demonstrated to be suitable for high-availability services and by putting a strong focus on robust and secure software during the entire development process.

Integrity is even more critical. Tampering with medical device instructions without anyone detecting this can have a severe impact on patient safety, which in turn would likely result in legal consequences for all involved parties. E.g., a deliberately wrong dosage of a substance, an altered configuration value of a life-sustaining machine or misinformation about allergens present in a product can lead to a serious endangerment of patients up to fatal complications. A lack of integrity protection can be exploited in a variety of attacks, e.g., by a disgruntled employee at the HCPM to harm their employer, by an external attacker to blackmail the HCPM or the dokspot service provider (request them pay a sum of money, otherwise tampering with the instructions will begin or continue), by a competitor of the HCPM, dokspot service provider or the HCI to gain an own advantage, and so on.

From this discussion, it follows that providing end-to-end integrity protection is crucial to achieve a truly trustworthy linking of physical products with online instructions. End-to-end integrity means that the HCP receives the original instruction that was provided by the HCPM, i.e., if the HCP receives and views an instruction on their device, then there is high assurance the instruction can be trusted and has not been tampered with. The bottom part of Figure 3 illustrates this end-to-end integrity: A HCPM employee provides an instruction and the dokspot service must make sure that the instruction is delivered to the HCP in the original form.

In the remainder of this paper, the focus will be on achieving this integrity-protection. Of course, other security aspects are relevant as well, but they can be solved by using state-of-the-art technologies and practices (as described, e.g., in [2]) and therefore do not require innovative approaches to be solved.

C. Threat Model

Figure 3 also includes possible threats against dokspot to compromise the integrity of instructions. The corresponding attack points are numbered 1–8 and explained in the following list.

- **Attack point 1**: A HCPM employee uploads manipulated instructions, either because they want to harm any of the involved parties or because they were bribed by someone else. Such internal attacks are common and their percentage among all cyber attacks is rising [3]. Note that the dokspot service provider could argue that this is out of scope: it is the problem of the HCPM to make sure no tampered instructions are uploaded in the first place. However, this would contradict the goal of being a truly trustworthy service, therefore dokspot should provide secure workflows that at least significantly increase the difficulty of such an insider attack.
- **Attack point 2**: The computer or web browser used by a HCPM employee to interact with the dokspot service could be compromised by an attacker. There are different attack vectors to achieve this, the most common ones include malicious e-mail attachments, drive-by downloads via compromised websites, and infected media (often USB sticks).
- **Attack points 3, 5, 7**: An attacker that has access to any of the communication channels can modify the transmitted instructions at will. This is a communication security problem and there exist good standard solutions to solve this problem by employing secure communication protocols, typically Transport Layer Security (TLS) [4]. Therefore, this attack point will not be addressed further in the remainder of this paper.
- **Attack point 4**: The attacker could get illegitimate access to the dokspot service, either by guessing or stealing (e.g., with a social engineering attack) the credentials of a HCPM employee or by directly compromising the service. While getting the credentials of users can effectively be prevented using strong authentication methods, it is much harder to make sure that the service cannot be compromised by exploiting a vulnerability. Putting a strong focus on secure software development can significantly reduce the risk of critical vulnerabilities, but today, there are no practical methods that can guarantee a 100% vulnerability-free service. The dokspot service primarily provides a web application interface, where security is especially hard to achieve: According to the Website Security Statistics Report of WhiteHat Security [5], 86% of several 10'000 analyzed websites contained at least one highly critical vulnerability. Therefore, the security architecture of dokspot should allow to guarantee integrity of instructions even if the service is (partly) compromised.
- **Attack point 6**: The storage service is another point of attack. While renowned companies are most likely taking great care with appropriate security measures, it is nevertheless possible that instructions can be manipulated while being stored in the storage service, either be external or internal attackers. Therefore, the

dokspot service should be able to cope with such attacks by making sure that modified instructions can be detected.

- **Attack point 8**: Just like with attack point 2, the computers, web browsers, or mobile devices used on the side of the HCI could also be compromised. This allows an attacker to exchange the requested instruction with any instruction the attacker wishes. In contrast to "standard" computers, mobile devices implement a stronger security model and therefore, attacks against mobile devices are much more difficult to execute and therefore occur significantly less frequently [15]. Most malware incidents on mobile devices happen because users install apps from untrusted sources. However, such malware is then confined to the actual app (due to the sandboxing model implemented by mobile devices) and can neither affect the underlying operating systems nor other apps (and also not the web browser). As a result, the risk of powerful malware on mobile devices that can affect the instruction that is requested and viewed in the web browser is small.

To summarize, there exist well-established security measures to secure (and integrity-protect) communications channels in the Internet. Therefore, attack points 3, 5, and 7 are marked with a green "solved mark" in Figure 3. In addition, mobile devices provide good protection against powerful attacks by design, so it is unlikely that they are compromised. With respect to all other attack points, however, there are no available standard solutions that could be applied.

III. Security Architecture

In this section, key components of the security architecture of dokspot are described. First, a secure workflow to manage online instructions is introduced. Next, the microservices-based approach of dokspot is described, followed by a brief introduction of the cloud infrastructure that is used by dokspot as a basis. After that, the usage of digital signatures to integrity-protect instructions during the entire lifecycle is explained before the section is completed by describing some further, more common security measures that are employed.

A. Secure Workflow to Publish Instructions

The goal of this workflow is to securely publish instructions and therefore make them publicly available to HCPs. To mitigate risks originating from a rogue HCPM employee (attack point 1), a compromised computer (attack point 2) or illegitimate access to the dokspot service (attack point 4), a workflow based on a segregation of duties (SoD) approach was introduced. The main idea is to split the process of publishing an instruction into several steps that are only executable by different authorized HCPM employees. To do this, Role-Based Access Control (RBAC) is used as the authorization mechanism [6] and the workflow enforces that at least three different roles must be involved in order to publish an instruction. As an employee can typically only have one of the three roles, this ensures that a minimum of three employees are required to publish an instruction. As a result of this, a single employee (and also two colluding employees) is never empowered to execute the entire workflow and is therefore unable to publish malicious instructions. Figure 4 illustrates the basic idea of this workflow.

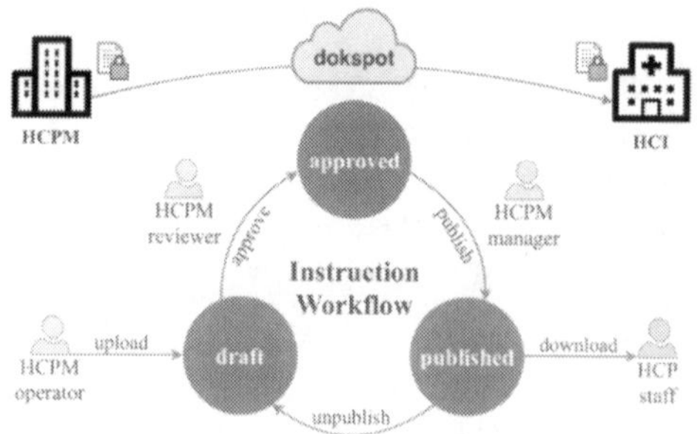

Figure 4. Workflow to Publish Instructions

As can be seen in Figure 4, an HCPM employee that has the role *operator* uploads an instruction to the dokspot service. As a result, the newly introduced instruction gets stored in state *draft* and needs to be approved by one or more HCPM employees with the role *reviewer*. A successful review process puts the instruction into the state *approved*. It can now be published by the HCPM manager of the corresponding healthcare product (corresponds to employees with the role *manager*). This results in a change of the state of the instruction to *published*, which means it is now publicly available and can be downloaded and viewed by the HCPs via the dokspot service.

This secure workflow prevents several attacks. For instance, if a malicious employee with role *operator* uploads a malicious instruction, this will most likely be detected by the employees with role *reviewer* as it is their obligation to review the content of the instructions for correctness. Likewise, it may be that malware on the computer of the *operator* modifies the instruction before it is uploaded, but just like in the first case, this should be detected by the *reviewers*. Note also that employees with roles *reviewer* and *manager* can only change the state of an instruction, but not its content, so it is not possible that one of them (or their computers, in case they are compromised) can modify an instruction in a malicious way.

B. Microservices

As mentioned earlier, the central component of the dokspot service is a web application. Nowadays, when developing a modern web application, developers heavily rely on pre-existing libraries. This includes frameworks that usually already consists of hundreds of thousands lines of code, plug-ins that themselves depend on many other plug-ins, middleware that helps to glue together all the pieces of a modern cloud-based architecture, and more. Unfortunately, the more complex a web application gets, the higher are the chances that one of its components has a security flaw that can be exploited by an attacker. That means that even if the developer of the web application itself is able to produce secure code (which is hard), there might still be some vulnerabilities lurking in the numerous dependencies, over which the developer has little control.

In addition, web applications often use monolithic architectures, which means there is usually one server (or multiple in the case of load-balancing) that is capable of handling all incoming requests. This implies that this server has to contain all of the required software components (which includes, as mentioned earlier, application, framework, plug-ins, middleware, etc.) at some point and must have full access to all data that is processed by the application. This leads to the unfavorable situation that attackers have many potential points of attack available to break into the system, and once they succeed in doing so, they are typically able to access and manipulate all data available to the web application.

To mitigate this risk, the dokspot service was split into multiple sub- or microservices, which handle just specific parts of the entire functionality. As a result of this, individual parts of the application, especially parts that handle sensitive data (e.g., login information), can be hardened against attacks, e.g., by blocking all requests outside their area of responsibility. Also, such a microservice requires just a small subset of the entire codebase and is therefore much harder to attack. Furthermore, the different microservices run on different servers with restricted access to the storage subsystems. This implies that if an attacker manages to get unauthorized access to one of the microservices, their possibilities are limited by the boundaries of the specific capabilities of the compromised service.

How to split up an application into microservices is dependent on the actual application. In the case of the dokspot service, this resulted in the microservices shown in Figure 5.

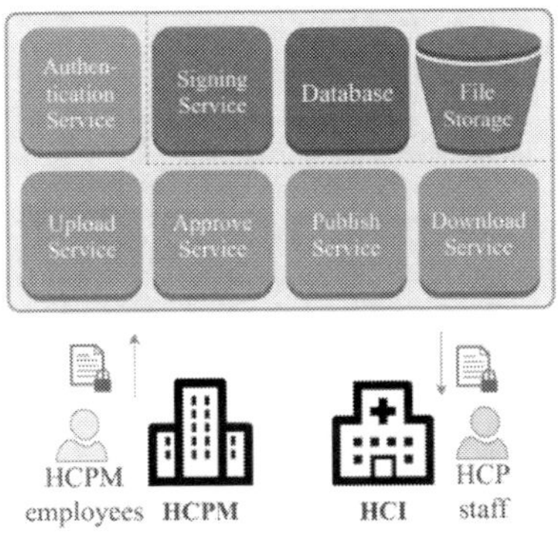

Figure 5. dokspot Microservices

Figure 5 shows the six microservices (in short *service*): authentication, upload, approve, publish, download and signing as well as the two storage types: database and file storage. Instructions are stored in the file storage, whereas all other data is stored in the database. The access privileges of the different services to the database and the file storage are restricted to the bare minimum, meaning that each service is only allowed to access or modify information which is required by the service in order to work as designed.

The authentication service is used to authenticate HCPM employees during login and to handle other tasks related to user management. The authentication service has exclusive access permissions to security-relevant data in the database such as the passwords of HCPM employees or their assigned roles. This also means that solely the authentication service is allowed to modify user accounts of HCPM employees and all modifications, such as password changes or role assignments, must happen via the authentication service. Naturally, due to its access to highly critical data, the authentication service is an attractive attack target, and therefore security was taken very seriously during its development. This included, e.g., working out a detailed security design as its basis, performing thorough code reviews and doing penetration tests. These are usually quiet costly activities, but due to the microservices-based architecture, they could be done in the context of a component with limited complexity, which allows to carry them out efficiently and which significantly increases the probability that the outcome is secure.

The upload, approve and publish services provide the functionality to perform the different steps during the workflow to publish instructions (as illustrated in Figure 4). They are only accessible by HCPM employees that have the corresponding role and only after they have successfully authenticated at the authentication service. This means the upload service is only accessible by authenticated HCPM operators, for the single purpose to upload instructions to dokspot. The upload service has exclusive rights to store instructions in the file storage and permissions to insert the corresponding metadata into the database. Once uploaded, an instruction can neither be modified nor deleted by the upload service or any other service, because the file storage does not allow it. Therefore, compromising any service does not allow an attacker to delete instructions. The approve service is used by authenticated HCPM reviewers and enables them to approve instructions after reviewing them carefully. The publish service is used by authenticated HCPM managers to make approved instructions publicly available.

The advantage of having different services to upload, approve and publish instructions becomes apparent under attack. For instance, a compromised upload service enables an attacker to upload malicious instructions with wrong or harmful content, but due to the service architecture, it is not possible to publish the instruction without compromising the approve and publish services as well. The reason is that the upload service lacks the capabilities and permissions on the database to complete the review or publish steps. On the other hand, if an attacker controls the publish service, it is still not possible to make harmful instructions publicly available, because the service is missing the upload functionality and permissions. To summarize, it is required to compromise all three services in order to publish an arbitrary instruction.

The download service is publicly accessible, without the need to authenticate, and is primarily used by HCPs. Nevertheless, the download service is critical, because if an attacker manages to compromise it, they can basically serve any instructions they like to the HCPs. Therefore, just like the authentication service, this service is also especially hardened and tested. As the service contains only relatively little functionality and only serves one single purpose, this was possible with a reasonable amount of resources.

The signing service is an internal service, which means it can only be accessed by some of the other services, but it is not accessible (and also not visible) from the Internet. This service is used to digitally sign instructions, which is explained in detail in Section III-D. The reason why this service is separated from the others and not publicly reachable is because it contains sensitive private keys used for signing, which must not fall into the hands of an attacker. If an attacker manages to get access to this key material, then they can publish fraudulent instructions, assuming they also get access to the database and the file storage. As an internal service, this service is considered difficult to compromise, because to start trying to attack it, an attacker first has to successfully compromise any of the other services. Or to put it differently: An attacker would be required to deeply infiltrate the dokspot service in order to reach the secrets to digitally sign instructions.

To summarize, the microservices-based architecture has several security benefits. First of all, the complexity of each service is much smaller compared to a monolithic approach, which reduces the attack surface of each individual service and which makes it easier to design, develop and configure them in a secure way. In addition, it reduces the impact of an attack, as in many cases, overall security is still maintained even if an attacker manages to compromise one of the services. And finally, it allows to hide services that provide functionality that must not be made available to the users (and therefore also the attackers), which further increases protection.

C. *Infrastructure*

More and more companies that provide Internet-based services do this over infrastructure of commercial cloud platform providers. As running an adequate data center usually is not a core competency of most companies, renting computing and network capacity is often the best option to meet the requirements at a reasonable cost. However, this leads to some loss of control over the service, which implies that it is crucial to pick a reputable provider that can demonstrate its trustworthiness, e.g., by possessing compliance certificates of advisable standards. If this advice is followed, one typically gets a higher level of security than by hosting the application and all data in a self-owned but unprofessionally managed infrastructure.

Dokspot is hosted on Amazon Web Services (AWS). The microservices are running on Heroku, which itself is hosted on AWS Elastic Compute Cloud (EC2). The database consists of several PostgreSQL instances provided by Amazons Relational Database Service (RDS). Every microservice has to authenticate against the database with role-specific credentials to restrict access to tables, columns and rows. This ensures that each microservice can only read and alter the smallest possible subset of data necessary for its role. The file storage is using an Amazons Simple Storage Service (S3) Bucket. To provide fast and reliable delivery CloudFlare is used as a Content Delivery Network (CDN) and Domain Name System (DNS) provider for all the microservices.

D. *Digital Signatures*

Usually, web applications guarantee the integrity of their data by carefully crafting the business logic in a way that does not allow for unwanted manipulation by the users. If there is a need to trace the changes that happen to data (e.g., to get an audit trail), some kind of logging mechanism is typically implemented. Unfortunately, there are two major weaknesses with this approach: a) One can never be sure that the code that handles the business logic is free from errors, and b) an administrator with sufficient access rights to the back-end of the application can often alter data and logs in an untraceable fashion. This weakens the guarantees one can make about the integrity of the data, which in the case of dokspot would go against one of the main goals. To mitigate the risks of such manipulation, dokspot uses digital signatures to strengthen the auditability of relevant actions. Based on this, unauthorized modifications can easily be detected.

Every relevant action executed by a dokspot user (operator, reviewer or manager) is digitally signed with a user-specific signing key. This is done with public key cryptography using RSA [7], but more modern signature schemes with smaller signature and key sizes, e.g., ECDSA [8] or ED25519 [9], could be used as well. The signed data covers metadata such as the user-id, the executed action, the digest of an uploaded instruction, etc. To make strong claims about the expressiveness of such signatures, it is absolutely crucial that only the dokspot users themselves have access to their own private key. In particular, this implies that even a service administrator or an attacker gaining access to one of the application servers cannot access the private keys of the users, as this would enable them to produce valid digital signatures in the users' names.

To achieve this, the private key of a user is stored in encrypted form in the database and this encryption uses a secret key that is derived from a user-chosen password. The dokspot service never sees this password and the private key is only decrypted and used to create signatures in-memory on the client-side (i.e., within the browser). Figure 6 illustrates how the key pair of a user is initially created.

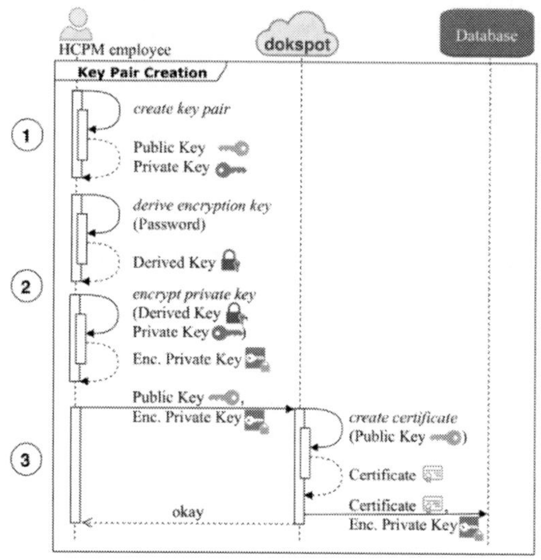

Figure 6. Key Pair Creation Sequence Diagram

In the first step (see Figure 6 (1)), after the first successful login, the user creates its own public/private key pair in the browser (using JavaScript code). Next (see Figure 6 (2)), they choose a dedicated password to protect the private key. To make brute forcing the password much more complicated in case an attacker gets access to an encrypted private key of a user, the Password-Based Key Derivation Function 2 (PBKDF2) [10] is utilized to derive a secret key based on the password. This secret key is then used to encrypt the private key. Once this has been done, the browser uploads the encrypted private key and the corresponding public key to dokspot (see Figure 6 (3)). Based on the received public key, the dokspot service then creates an X.509 certificate [11] and as a result of this, the key pair is now certified and can be used for signing and for verification of corresponding signatures in the context of the dokspot service. Finally, both the certificate and the encrypted private key are stored in the database.

Once this has been completed, the user can create signatures, e.g., to sign instructions during the upload step. This process is illustrated in Figure 7.

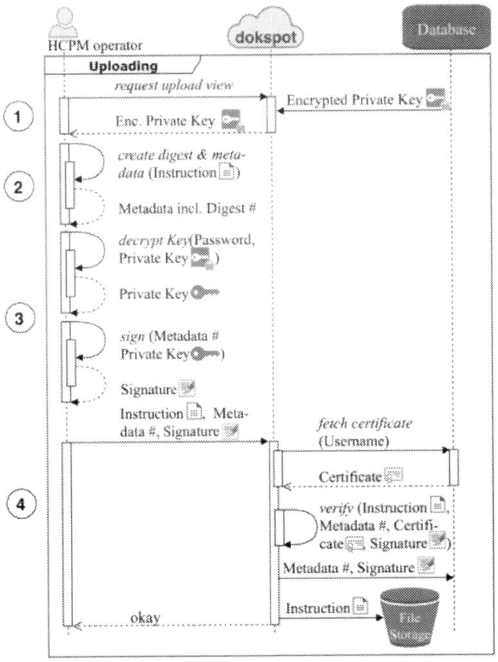

Figure 7. Instruction Uploading Sequence Diagram

To upload an instruction, the user first navigates to the corresponding view of the dokspot service by using their web browser. Before sending the web page to the browser, the dokspot service fetches the user's encrypted private key from the database and embeds it into the page (see Figure 7 (1)). Next, the user picks the instruction they want to upload and enters the associated metadata (title, language, etc.) (see Figure 7 (2)). In the background, a JavaScript function starts calculating the digest for the chosen instruction, using a cryptographic hash function. Once the user has finished entering the metadata, they start the signing process (see Figure 7 (3)) by clicking a button. The user is asked to enter the password that was used to protect their signing key and if the password is correct, the private key is decrypted. A JavaScript routine then adds a timestamp and the digest to the metadata and signs the resulting metadata with the private key.

Next, the instruction, the metadata, and the signature are sent to the dokspot service (see Figure 7 (4)). Upon receiving the data, the service fetches the certificate of the current user from the database and uses the public key in the certificate to verify the signature. If it is correct, it stores the instruction in the file storage and the metadata including the signature in the database. As a result of this, the signature now seals both the instruction and the metadata. If an attacker manages to alter just a single bit in this instruction later during the lifecycle of the instruction, this can easily be detected as verifying the signature will fail.

When a user with the appropriate role changes the state of an instruction (e.g., from *draft* to *approved*, see Figure 4) then this action also results in creating a digital signature. Technically, this works similar as as in Figure 7, meaning that the user navigates to the corresponding view, which again includes the user's encrypted private key. As soon as the user wants to initiate the state change (e.g., after they have checked the instruction and made sure it can be approved), they click a button, which triggers the change of the state: In a first step, further metadata is produced, which includes the specific action to be performed (e.g., the approval of the instruction) and a timestamp. Next, the digest of the instruction is included in the metadata and the resulting metadata is signed with the user's private key (which, just like earlier, requires the user to enter their password). All of this is then sent to the dokspot service, which checks if the user is allowed to execute the specified action and which checks the validity of the signature. If all of the requirements are satisfied, the received metadata is stored in the database and the state of the instruction is updated in the database.

To further enhance the auditability, the dokspot service adds its own signatures to important actions. So when the service receives an uploaded instruction or a state change, a second signature, which acknowledges the reception, is created and saved along with the metadata and the signature of the user. This additional signature serves as a proof that the dokspot service verified the user's signature of the action and that is has been declared valid.

When a HCP wants to access an instruction, they have to navigate to the appropriate view, which shows the published instructions for a specific product. After the user picks the instruction they want to read, the dokspot service collects all the signed metadata attributed to this instruction. The service then checks if all the necessary signatures are available and valid, i.e., it is checked whether the instruction contains valid signatures to verify it was uploaded, it was approved, it was published and so on. To do these checks, the service freshly calculates the digest of the requested instruction and cross-checks it with the digest specified in the signed metadata. If any of the checks fail, the instruction will not be delivered to the HCP as this is an indication of an attack.

E. *Passwordless Signatures*

So far, it is assumed the users use a dedicated password to protect their private key. This mitigates a range of attacks, as this password will never be sent across the network. However, it requires that the user has to remember a second password (in addition to the login password). Also, they have to enter the password that protects the private key every time a signature should be created. While this is the most secure configuration, it is also somewhat inconvenient. For this reason, a passwordless signing process was developed, which lets a user trade some of the security for a more convenient experience of the dokspot service.

With passwordless signing, the login password is used to encrypt the private key. After the user has entered this password during login (which they have to do anyway when using dokspot), the password is temporarily stored so it can be used to automatically decrypt the private key at a later point. Storing the password in plain text on the client side would not be ideal from a security perspective as it may provide an attacker that somehow gets access to the system with an opportunity to extract this password. Therefore, the password is protected with the approach illustrated in Figure 8.

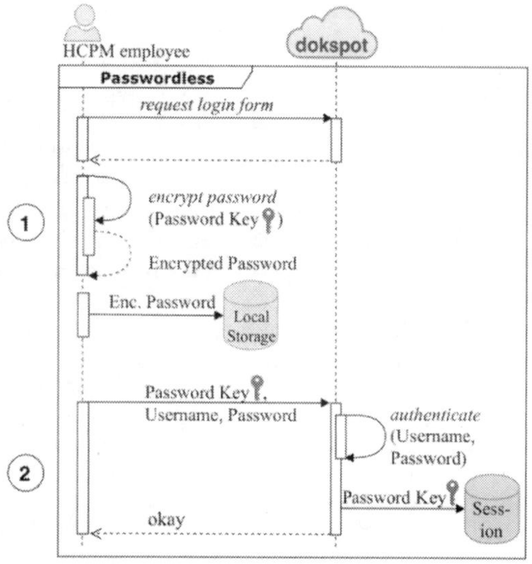

Figure 8. Passwordless Signing Sequence Diagram

When the user opens the login page, a JavaScript function in the background choses a sufficiently large symmetric key at random, identified as the password key (see Figure 8 (1)). This password key is stored in a hidden field in the login form. When the user sends the login form with the password key to the dokspot service, a JavaScript function in the browser encrypts the entered password with the password key and stores the encrypted password in the browsers local storage. If the authentication is successful, the service stores the received password key in the session of the user (see Figure 8 (2)). This means that once the login is done, both sides hold just one piece of the information needed to decrypt the user's password. The client needs the password key from the service to decrypt the locally stored encrypted password and the service would need the encrypted password to do the same. That means if an attacker gains access to only the client (or only the service), they will not be able to retrieve the password.

Note that during a login procedure with a simple POST request, the user's password must be exposed to the dokspot service. This implies that theoretically, the service could try to decrypt the user's private key at that point. This can be prevented by using a more secure login scheme, utilizing, e.g., the Secure Remote Password (SRP) Protocol [12].

The process of creating a passwordless signature is essentially the same as described earlier in Figure 7. This time, however, the user does not only get the encrypted private key from the dokspot service, but additionally also the symmetric password key. Next, a JavaScript function in the browser will use the password key to decrypt the encrypted password that is stored in the local storage. And finally, the decrypted password can be used to decrypt the private key, which can then be used to create digital signatures.

The confidentiality of the user's password is not affected by this signing scheme. Independent of entering the password manually or recovering it via JavaScript, the password is accessible in the JavaScript runtime environment during the signing process either way.

F. Common Security Measures

Besides the very specific and innovative security measures described earlier, the dokspot service also employs several state-of-the-art security measures, some of which are briefly summarized in this paragraph. First of all, the dokspot service can only be reached using HTTP over TLS (HTTPS). This provides an encrypted and integrity-protected communication channel. An HTTP Strict Transport Security (HSTS) policy is in place and preloaded, to further increase the difficulty of an attack against the connection. The service uses secure cookies (encrypted and signed), a Content Security Policy (CSP) and cross-site scripting protection to reduce the risk of a break-in. The DNS entries of the dokspot domain are protected by DNSSEC [13] and DNS Certification Authority Authorization (CAA) to make it difficult for an attacker to reroute users to a fake service. The login, located on the authentication service, can be secured using two-factor authentication. This mitigates the risks of a stolen, lost or phished password. In addition, some HCPM employees (e.g., the manager of an entire product family), can be notified if a suspicious or important events, such as publishing an instruction to the public, takes place. This allows to quickly react in case of a potential security breach. Finally, to verify the high level of security, the dokspot service has been tested for vulnerabilities by security professionals.

IV. Evaluation

Coming back to the threat model in Section II-C, we can see that the risks in the context of the identified attack points could be mitigated, except for attack point 8.

If, due to attack points 1 or 2, a non-genuine instruction gets uploaded onto the dokspot service, this cannot be detected automatically because the platform itself is oblivious of the content of uploaded instructions. The workflow, however, enforces an instruction to be reviewed by multiple parties before it can be published to the public audience. Assuming that at least one of the involved parties performs their part of the process diligently, a forged instruction will never be made accessible to the HCPs.

To mitigate the risk of a break-in into one of the servers, as mentioned in attack point 4, the service is split into multiple microservices and all relevant actions are digitally signed. The microservices that handle the processes of uploading, approving and publishing instructions are the ones with the widest range of functionality. They are therefore the most probable candidates to contain vulnerabilities that can be abused to gain illegitimate access to the dokspot service. However, an attacker with access to any subset of these exposed microservices will not be able to produce the full valid set of signatures required for an instruction in state *published*. As the download service checks the full set of signatures before an instruction is delivered to an HCP, it will not be possible for an attacker to provide manipulated instructions through dokspot.

An exposed point in this scenario is the download microservice. An attacker controlling this component can deliver whatever instructions they wish to HCPs because it allows the attacker to bypass any form of signature validation. But as the download microservice just offers a single and very simple functionality, it is relatively easy to harden this service to a point where a successful break-in is very unlikely.

In attack point 6, a breach of the storage system is assumed. As Amazon RDS and S3 are used for storage, we will not discuss how an attacker might achieve this and focus on the consequences. The signed metadata described in Section III-D holds, among other information, the digest of the corresponding instruction. Dokspot regularly verifies that the digest of the instructions matches the digests that are stored as part of the signed metadata. This means that if any instruction – even a single letter – were modified for whatever reason, this would be detected during such a check. Also, as mentioned earlier, the full set of signatures is checked whenever an instruction is requested by an HCP, where any manipulation would be detected as well. To summarize, this means that any kind of breach of the storage system cannot result in delivering a manipulated instruction to the HCPs.

Unfortunately, attack point 8 remains an open problem. Assuming the attacker has control over the device used by an HCP to download an instruction, there is nothing the dokspot service can do to prevent the attacker from displaying whatever instruction they wish. Therefore, preventing such attacks is currently out of scope for dokspot and it is the responsibility of the HCI to make sure its IT infrastructure is malware-free.

V. Related Work

Microservices recently became popular as a service architecture in the web environment [14] and provide, compared to a monolithic approach, benefits, such as scalability, cost reduction and improved performance [15]. While some papers discuss the security challenges of microservices, such as authentication or communication between services [16], as well as auditability or inter-service trust [17], they do not highlight the security benefits. Therefore, using microservices to increase application security – as done with the dokspot service – appears to be a novel approach.

The idea of using digital signatures to safeguard business workflows has been covered in multiple papers since 1999 (e.g., [18], [19]). However, previous works do not cover the particular characteristics that must be considered when using signatures in modern web applications or role-based microservices. In the domain of user signatures, Halpin wrote a critical acclaim of the W3C web cryptography API (and browser-based cryptography in general) [20], which highlights some limitations that come with the fact that the web server is mostly in control over the executed JavaScript code. To address this, our approach with SoD for HCPM employees and microservices makes it very difficult for an attacker to gain control over the necessary key material to create a valid set of signatures.

VI. Conclusion

In this paper, we presented dokspot, which is an Internet-based service to securely link physical products with online instructions. To achieve this, dokspot is based on a sophisticated security architecture that combines several approaches. We have shown that with the help of a well-tailored workflow, strong claims can be made about the genuineness of instructions managed by the dokspot service. By utilizing digital signatures on multiple layers, the compliance of the workflow can be proven cryptographically and the integrity of instructions can be guaranteed. By appropriately splitting the service into microservices, the dokspot service gets hardened as a whole. Due to restrictions with respect to the access permissions of the microservices and the use of digital signatures, the possibilities for an attacker are greatly restrained, even in the case of a successful partial break-in. Finally, we introduced a passwordless signature scheme, which leads to a more convenient user experience when creating digital signatures without significantly reducing security. Overall, the presented security architecture makes dokspot highly resistant to a wide range of attacks.

Acknowledgement

This work was partly funded by the Swiss Confederation's innovation promotion agency CTI (project 18990.1 PFES-ES).

References

[1] MHRA, "GXP Data Integrity Guidance and Definitions," MHRA, Revision 1, Mar. 2018.

[2] OWASP, "OWASP Guide Project," 2014. [Online]. Available: https://www.owasp.org/index.php/OWASP_Guide_Project 2018.08.08.

[3] D. M. Upton and S. Creese, "The Danger from Within." *Harvard business review*, vol. 92, no. 9, pp. 94–101, 2014.

[4] T. Dierks and E. Rescorla, "The Transport Layer Security (TLS) Protocol Version 1.2," Internet Requests for Comments, RFC Editor, RFC 5246, Aug. 2008. [Online]. Available: https://www.rfc-editor.org/rfc/rfc5246.txt 2018.08.08.

[5] WhiteHat Security, "Website Security Statistics Report," May 2013. [Online]. Available: https://www.whitehatsec.com/wp-content/uploads/2013/05/WPstatsReport_052013.pdf 2018.08.08.

[6] R. S. Sandhu, E. J. Coyne, H. L. Feinstein, and C. E. Youman, "Role-Based Access Control Models," *Computer*, vol. 29, no. 2, pp. 38–47, Feb. 1996.

[7] K. Moriarty, B. Kaliski, J. Jonsson, and A. Rusch, "PKCS #1: RSA Cryptography Specifications Version 2.2," Internet Requests for Comments, RFC Editor, RFC 8017, Nov. 2016. [Online]. Available: https://www.rfc-editor.org/rfc/rfc8017.txt 2018.08.08.

[8] D. Johnson, A. Menezes, and S. Vanstone, "The Elliptic Curve Digital Signature Algorithm (ECDSA)," *International Journal of Information Security*, vol. 1, no. 1, pp. 36–63, 2001.

[9] D. J. Bernstein, N. Duif, T. Lange, P. Schwabe, and B.-Y. Yang, "High-Speed High-Security Signatures," *JCEN*, vol. 2, no. 2, pp. 77–89, 2012.

[10] B. Kaliski, "PKCS #5: Password-Based Cryptography Specification Version 2.0," Internet Requests for Comments, RFC Editor, RFC 2898, Sep. 2000. [Online]. Available: https://www.rfc-editor.org/rfc/rfc2898.txt 2018.08.08.

[11] D. Cooper, S. Santesson, S. Farrell, S. Boeyen, R. Housley, and W. Polk, "Internet X.509 Public Key Infrastructure Certificate and Certificate Revocation List (CRL) Profile," Internet Requests for Comments, RFC Editor, RFC 5280, May 2008. [Online]. Available: https://www.rfc-editor.org/rfc/rfc5280.txt 2018.08.08.

[12] T. D. Wu, "The Secure Remote Password Protocol." in *NDSS*, vol. 98, San Diego, CA, USA, Mar. 1998, pp. 97–111.

[13] M. Larson, D. Massey, S. Rose, R. Arends, and R. Austein, "DNS Security Introduction and Requirements," Internet Requests for Comments, RFC Editor, RFC 4022, Mar. 2005. [Online]. Available: https://www.rfc-editor.org/rfc/rfc4033.txt 2018.08.08.

[14] J. Thönes, "Microservices," *IEEE Software*, vol. 32, no. 1, pp. 116–116, Jan. 2015.

[15] M. Villamizar, O. Garcés, H. Castro, M. Verano, L. Salamanca, R. Casallas, and S. Gil, "Evaluating the Monolithic and the Microservice Architecture Pattern to Deploy Web Applications in the Cloud," in *2015 10th Computing Colombian Conference (10CCC)*, Sep. 2015, pp. 583–590.

[16] N. Alshuqayran, N. Ali, and R. Evans, "A Systematic Mapping Study in Microservice Architecture," in *2016 IEEE 9th International Conference on Service-Oriented Computing and Applications (SOCA)*, Nov. 2016, pp. 44–51.

[17] Y. Sun, S. Nanda, and T. Jaeger, "Security-as-a-Service for Microservices-Based Cloud Applications," in *2015 IEEE 7th International Conference on Cloud Computing Technology and Science (CloudCom)*, Nov. 2015, pp. 50–57.

[18] K. R. P. H. Leung and L. C. K. Hui, "Signature Management in Workflow Systems," in *23rd International Computer Software and Applications Conference*. Washington, DC, USA: IEEE Computer Society, Oct. 1999, pp. 424–429.

[19] H. W. Lim, F. Kerschbaum, and H. Wang, "Workflow Signatures for Business Process Compliance," *IEEE Trans. Dependable Secur. Comput.*, vol. 9, no. 5, pp. 756–769, Sep. 2012.

[20] H. Halpin, "The W3C Web Cryptography API: Motivation and Overview," in *Proceedings of the 23rd International Conference on World Wide Web*. Seoul, Korea: ACM, 2014, pp. 959–964.

Interactive Exploration and Querying of RDF Data

Martin Kryl[12], Petr Vcelak[12] and Jana Kleckova[12]

[1]NTIS - New Technologies for the Information Society, University of West Bohemia, Czech Republic
[2]Department of Computer Science and Engineering, University of West Bohemia, Czech Republic
email: {kryl, vcelak, kleckova}@kiv.zcu.cz

Abstract—**Many contemporary healthcare information systems incorporate and utilize Resource Description Framework (RDF) datasets, which are characteristic by their flexibility and ability to form complex data networks. Users might find themselves overwhelmed when trying to understand the data layout since there are no apparent rigid structures such as tables in relational databases. In this paper, we present a prototype data exploration tool, that enables users to grasp the data structure by exploring a simplified RDF model. The solution does not rely on ontological description. The visualization has four modes of interaction defined that allow exploration in different levels of detail. One of the modes can be used to interactively create a SPARQL SELECT query. The proposed solution combines graph visualization and data extraction techniques into a single tool and allows users without expert SPARQL knowledge to extract data from RDF graph.**

Keywords-RDF; visualization; data exploration; interactive SPARQL builder.

I. Introduction

The current trend in data storage is to use various nonrelational databases (NoSQL) and data models beside classical relational databases. Big Data are common in health domain and relational databases are not considered suitable for dealing with them since they lack horizontal scalability, and need hard consistency [1]. One of the currently popular NoSQL data models is RDF, which has found its use as the preferred model for Open Data [2] and it is also used in Medical Information Systems for its ability to integrate heterogeneous data. Medical RDF applications can range from custom prospective study databases [3] to systems for inter-hospital data exchange [4]. RDF model is directed multigraph, and can be queried by special query languages, e.g., SPARQL.

There are some disadvantages of keeping data in RDF model. The main issue seems to be the lack of RDF support in common analytical and Business Intelligence software. Users usually need to transform the data into tabular format before being able to do the analytics, which requires additional knowledge of RDF query language. Another issue arises when data analysts want to understand the content of a dataset by performing data exploration. The structure of raw data is complex and there is only limited support in some graph database systems for navigating through the graph [5]. Various semantic browsers for traversing RDF graphs exist, such as Tabulator [6], but it can be difficult to use them for large graphs. The user can only see immediate surrounding nodes and may get lost during graph traversal. One can try to visualize the dataset but with the growing number of resource nodes, the legibility of visualization quickly decreases. Providing a method for interactive RDF analytics, that would not require the user to manually write queries, is currently considered to be an important area of research [5].

We have been working on a prototype solution allowing users to explore a general RDF data model and interactively define a data projection above the dataset without the need of extensive knowledge of RDF and query languages. The solution is composed of three components. The first one is an RDF model crawler, which analyses the model structure, determines property cardinalities and prevalence of RDF types. The second one is a web visualization which utilizes findings of the crawler to provide an aggregate graph view with a possibility of interactive model exploration. The third is a query builder that provides auto-generated SPARQL queries based on the user interactions with the visualization. The user can select objects of interests in the visualization and get the transformed underlying data.

There are multiple related projects that either help users to build SPARQL queries [7][8] or provide an aggregated visualization of data model [9]. However, to the best of our knowledge, there are no solutions that would assist in data exploration and extraction by combining the two techniques.

Both of the query builders work with a fixed set of SPARQL endpoints and have no visual tool that would help the user understand the relations in dataset. SPARQL Builder [7] constructs the query in two steps. First, the user selects two RDF classes from the dataset. The application prints all the possible paths between the resources of selected classes and the user has to choose one of them forming the query pattern. SPARQLGraph [8] offers an intuitive drag & drop query builder allowing the modeling of more complex query patterns than SPARQL Builder. The visualization [9] uses aggregated model based on the RDF classes of resources and provides good overview of relations in the dataset. However, it is not possible to use the visualization to interactively generate a query.

In the following Section II, the structure and functionality of the proposed system is described. The prototype implementation details are provided in Section III. The achieved results are discussed in context of other related solutions in Section IV before the conclusion and future outlook in Section V.

II. Proposed System

The system is composed of data preprocessor, graph visualization and query builder. Every RDF dataset needs to be transformed into corresponding aggregated model first. The aggregated model describes general structure and relations found in the dataset and serves similar purpose as widely accepted Entity-Relational model in relational databases. Visualization draws the model and allows user to interact with it. Query builder allows selection and projection of source data.

A. Data preprocessing

RDF data utilize property *rdf:type* to indicate that the resource is an instance of the specified class. Knowing the classes of resources, aggregated model is formed by finding all distinct combination of (Sc, P, Oc), where Sc is the class of a subject resource, P is the property and Oc is the class of an object resource. Resources having no class defined are ignored at this time. Literal values of datatype properties are considered to be belonging to a pseudo-class with the same id as the property, i.e., (Sc, P, P). In theory, it would be possible to generalize this aggregation by choosing arbitrary property instead of *rdf:type*, but in practice, there seems to be no property with such a prevalence.

Additionally, cardinalities of all properties in respect to subject and object classes are calculated, i.e., number of different values of type Oc are counted for each instance of Sc in (Sc, P, Oc). Visualization uses information about minimum and maximum cardinalities, as well as histogram of cardinality values. Lastly, total number of instances of each class are calculated.

B. Visualization

Metadata collected in previous step are incorporated into interactive directed graph visualization. Nodes represent resource classes or literal pseudo-classes, while edges represent RDF properties. Color of the node is used to distinguish between regular and pseudo-classes. Existence of an edge labeled P directed from A to B means that there is at least one RDF triple in the original data, where an instance of A is related to an instance of B by property P. Edges are visualized as arcs with clockwise orientation instead of lines. This way inverse properties do not overlap and it is also possible to visualize multiple edges in the same direction by assigning different radius for each arc. By using this technique, it is possible to draw loop edges the same way as other edges. Styling of the edge indicates the minimum and maximum cardinality of the property in context of classes A and B:

- Solid arc represents the minimum cardinality of 1, i.e., each instance of A has at least one value of class B assigned by the property.
- Dashed arc indicates minimum cardinality of 0, i.e., there are instances of A that have no instance of B assigned by the property.
- Empty arrow marker means that the maximum cardinality of the property is 1, i.e., there is always at most one value of B.
- Filled arrow marker means that the maximum cardinality is greater than 1, i.e., the property may have multiple values.

Following the Visual Information Seeking Mantra: 'overview first, zoom and filter, then details-on-demand' [10], the visualization offers four modes of interaction.

Overview: By default, the visualization shows only nodes representing RDF classes and object properties between them. This is the minimal possible configuration showing the relations in the entire dataset. Only the local names of a resources, the last part of object URIs (Uniform Resource Identifier), are displayed, unless there is an ambiguity. The user can scroll through the visualization canvas and move individual nodes. If there are multiple properties with the same orientation between any two resources, they can be merged into a single arc with numerical label indicating the number of merged properties

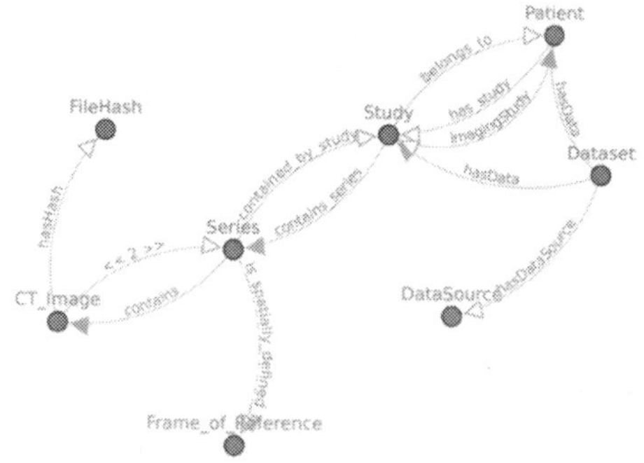

Figure 1. Visualization of an aggregated model of medical imaging data consisting of 8 RDF types and showcasing all possible cardinality types. Literal nodes and datatype properties are hidden.

as can be seen in Figure 1 between nodes labeled *CT_Image* and *Series*, or they can be drawn as separate arcs as shown in Figure 1 between *Patient* and *Study* nodes.

Zoom: This mode allows an expansion of detail in an area of interest. By clicking on the node, the user can choose to display datatype property edges leading from the specified class to respective pseudo-class nodes. Merged properties can be expanded or merged back by clicking on the arc.

Filter: The user can filter displayed graph elements by selecting classes of interest via a set of checkboxes provided next to the visualization canvas. All pseudo-classes are tied to a single checkbox. Only the nodes of selected classes and the edges between them are displayed. Additionally, one can set all adjacent nodes of a given node to be visible by clicking on the node and thus updating the filter.

Details-on-demand: An infobox containing information about the selected property prevalence and histogram of its cardinalities is displayed next to the graph visualization as shown in Figure 2a. It is also possible to interactively query the original dataset in this mode. The user can highlight nodes and edges that will serve as a subgraph pattern in the SPARQL query. The underlying query string is displayed and dynamically updated as the user interacts with the visualization. Details on SPARQL creation are provided in a further section. Some segments of the generated SPARQL query might be highlighted in some cases as seen in Figure 2b. The user can click on the text and tweak the query by selecting alternative auto-generated segment. Results are displayed in tabular form under the visualization and can be exported to a CSV (comma-separated values) file.

C. SPARQL query builder

Query builder works with an active selection in the visualization. Variables in the SPARQL query use the same name as the nodes they represent, thus the same variable name is always used for one node. Highlighted subgraph edges and nodes are collected and the first node to be traversed is chosen. The subgraph is traversed by depth-first approach with edges of datatype properties having priority. For each traversed edge, a new triple is added to the WHERE clause declaring the relation between the two variables via URI of the edge. A triple definition of variable type is added if the *rdf:type* of the variable has not been defined before and the variable is not literal. This default behavior works well when using only properties with cardinality of 1.

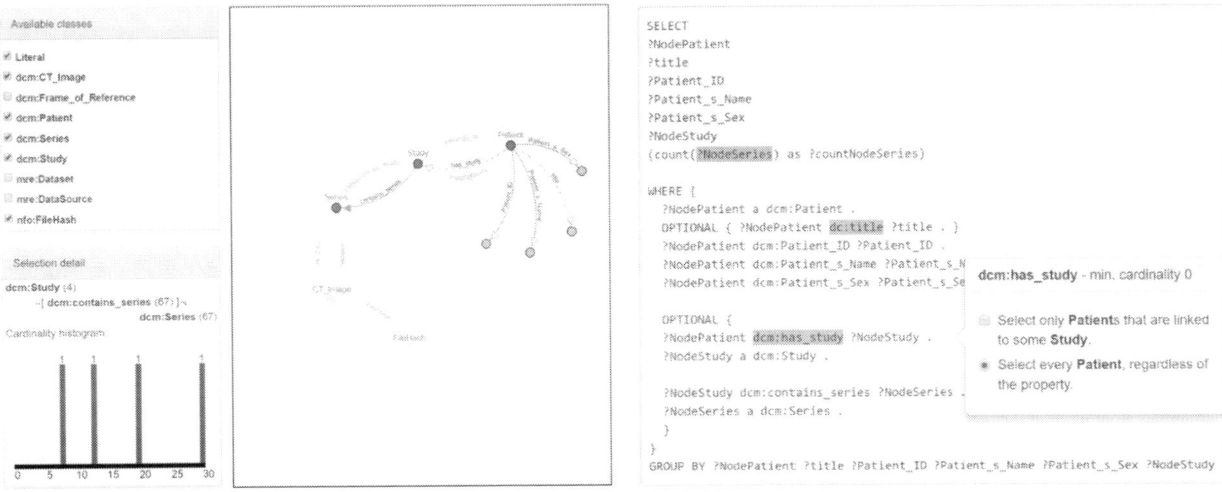

(a) (b)

Figure 2. Interactive query generation. Selection detail box in lower left part of (a) provides information about a single graph edge. Numbers in brackets indicate the prevalence in the dataset. The histogram shows cardinality spectrum of the property. The query in (b) is generated based on the highlighted subgraph from (a). There are three highlighted opportunities to tweak the query via popover box.

When dealing with properties having minimum cardinality of 0, the part of SPARQL query is highlighted, thus visually prompting the user to make a decision on a way this property should be handled. Without further changes, the property is considered mandatory and any resource that does not have the property assigned will not match the pattern and will not be present in the result. An alternative to this is to wrap the triple with OPTIONAL clause. This is sufficient for datatype properties, but for object properties it might lead to cartesian product evaluation in case the potentially unbound variable is used further in the query. To prevent this behavior, entire query pattern generated by traversing the target node and its children is wrapped with the optional clause.

All the variables declared in pattern statements are inserted into SELECT clause of the query. Clicking on the variable name will insert a simple filter at the end of the where block, which could be further edited by the user. This allows for additional conditioned selection in the query, thus limiting the result space.

Query builder displays a warning text in two specific situations. First, if the current selection does not form a pattern of connected graph. Proceeding with the query execution would result in a cartesian product of the disjoint components, which is undesirable in most cases. There is no clear-cut solution to this problem and the user needs to change the subgraph selection manually. The second case is when multiple edges between two nodes are selected. Only instances, that has all the properties leading to the same instance will be in the resulting selection. This is usually not the behavior the user wants, but it might be desirable in some cases.

III. Implementation

The prototype is implemented using client-server architecture. Server part is written in Java using Spring Framework [11] for web communication and Apache Jena [12] for RDF manipulation. The server accepts either an RDF file in serialization format supported by Jena or an URL address of SPARQL endpoint that implements the SPARQL 1.1 Graph Store Protocol [13]. Currently, the server application analyses the contents of the dataset programatically via Jena API. This provides parallel processing support of the input model if needed in the future. The chosen approach is in contrast to other contemporary solutions, which gains the graph metadata purely by analytically querying of public SPARQL endpoints built upon the datasets. Our metadata collection workflow could be transformed to a batch of analytical queries as Jena framework enables to query loaded RDF models, as well as remote SPARQL endpoints. The resulting dataset description metadata are returned as a JavaScript Object Notation (JSON) file. The server also provides an interface for SPARQL SELECT query execution above the dataset.

The client is written in HTML5 and utilizes D3 JavaScript library [14] to create a force-directed graph based interactive visualization. The visualization uses metadata JSON file provided by the server. The client is platform independent and only requires modern web browser with enabled JavaScript support.

IV. Discussion

The main focus of this project was on creating a tool that would assist the user in filtering and transforming the RDF dataset into tabular data structure for further use in other analytical software. The visualization interactions were inspired by the visual interface of Microsoft Access Query Designer, which users can successfully operate with only a basic understanding of relational database theory. Entity-Relationship model is the fundamental diagram in relational databases to understand the structure of dataset and Query Designer uses it to define joins between tables. In RDF case, we have used *rdf:type* values of resources as the entity labels and constructed an aggregate graph explaining the relations present in the dataset between such entities. By highlighting segments of the graph, the user intuitively defines a query projection.

There are minimum and maximum cardinalities calculated in preprocessing phase, which are used during query generation to determinate the need for optional clause and aggregation definitions. These values could in theory be gained from ontological definitions of properties, however we would have to accept the assumptions that the ontologies are available, the ontological restrictions are defined and data respect the ontologies. To allow the user to work with datasets lacking the ontological description, we have decided to use the ontologies only as a supplementary source of annotations. Similarly, the entire aggregate graph could in theory be created based on the ontological definitions of classes, properties and their domain and range spaces, but again we have chosen to rather describe the exact state of data as presented in an input file.

The prototype solution uses a generic approach that is applicable for any RDF file or SPARQL endpoint allowing access via Graph Store Protocol. Users are thus able to explore and extract data from various sources and are not limited to predefined databases, which is a common issue in other SPARQL builder solutions [7][8]. To support on-demand viewing of data and export of projected table, the server needs to store the input datasets. Currently, the entire datasets are read into Jena in-memory model which is a major scalability concern for future development because Java Virtual Machine memory capacity is limited.

Preprocessing of the dataset is computationally expensive operation and it scales linearly with the triple count. Depending on the input dataset size, there might be a noticeable delay before the user can interact with the visualization and it might be worth searching for possibilities on how to show results of partially processed dataset to the user. The visualization [9] avoids the problem by running the process in regular interval and having a cached results available on-demand. This is not applicable in our case, since user can input arbitrary datasets.

The quality and clarity of visualization depends mainly on two factors, the number of different RDF types and their hierarchy. With increasing node count, the visualization spans larger area and it becomes harder for users to grasp the overall shape or even find property paths between the nodes of interest. In such cases, it might be useful to incorporate a filter method that would only display the two selected nodes and the property path between them similarly as SPARQL Builder [7] does. User could then expand the neighboring nodes and the two nodes would serve as starting point for further exploration. The problem with hierarchy could be partially alleviated by using the drill down and drill up operations on the hierarchy of RDF types as was done in [9]. However, this would conceal some of the types and a user not knowing about the subtypes and supertypes of searched term might be confused.

V. Conclusion and Future Work

We have introduced the prototype software solution for interactive RDF data exploration and transformation to tabular format in this paper. The proposed interactions are in accordance with the overview first, zoom and filter, then details-on-demand principle. SPARQL generator is a part of the solution and is used for building basic SELECT queries with support of optional blocks, filter clause and group by clause. The user is prompted to choose from several offered query snippets rather than freely edit the query string, and is not expected to know the query language in detail.

The solution currently works well for datasets containing low number of RDF classes, but it needs to be improved before deploying and integrating in other systems. The solution might assist doctors in extracting and transforming relevant data for their clinical research or it might help in initial orientation in data structure of information system. It might find its use in applications for manipulating RDF data, e.g., an RDF editor.

Future work consists of extending the system architecture to include a proper database layer to store the input datasets and metadata generated during preprocessing. We will also look into the ways to optimize the preprocessing phase and to gather the model metadata by a batch of analytical SPARQL queries, thus allowing the visualization to be run against SPARQL endpoints built upon the datasets. Later on, we would like to perform usability testing.

Acknowledgment

The work was supported from European Regional Development Fund-Project "Application of Modern Technologies in Medicine and Industry" (No. CZ.02.1.01/0.0/0.0/17_048/0007280) and by Ministry of Education, Youth and Sports of the Czech Republic, project PUNTIS (LO1506) under the program NPU I and project SGS-2016-018.

References

[1] I. Merelli, H. Pérez-Sánchez, S. Gesing, and D. DAgostino, "Managing, Analysing, and Integrating Big Data in Medical Bioinformatics: Open Problems and Future Perspectives," BioMed research international, vol. 2014, 2014.

[2] T. Berners-Lee, "Linked Data - Design Issues," 2010, URL: https://www.w3.org/DesignIssues/LinkedData.html [retrieved: August, 2018].

[3] J. Schwarz et al., "Inflammatory bowel disease incidence in Czech children: A regional prospective study, 2000-2015," World journal of gastroenterology, vol. 23, no. 22, 2017, p. 4090.

[4] P. Rajbhandari, R. Gosai, R. C. Shah, and K. Pramod, "Semantic Web in Medical Information Systems," International Journal of Advances in Engineering & Technology, vol. 5, no. 1, 2012, p. 536.

[5] M. Junghanns, A. Petermann, M. Neumann, and E. Rahm, "Management and Analysis of Big Graph Data: Current Systems and Open Challenges," in Handbook of Big Data Technologies. Springer, 2017, pp. 457–505.

[6] Berners-Lee et al., "Tabulator: Exploring and Analyzing Linked Data on the Semantic Web," in Proceedings of the 3rd International Semantic Web User Interaction Workshop, vol. 2006. Citeseer, 2006, p. 159.

[7] A. Yamaguchi et al., "SPARQL Builder: Constructing SPARQL Query by Traversing Class-Class Relationships for Life Science Databases," in JIST (Workshops & Posters), 2016, pp. 58–61.

[8] D. Schweiger, Z. Trajanoski, and S. Pabinger, "SPARQLGraph: a web-based platform for graphically querying biological Semantic Web databases," BMC bioinformatics, vol. 15, no. 1, 2014, p. 279.

[9] F. Florenzano, D. Parra, J. Reutter, and F. Venegas, "An Interactive Visualisation for RDF Data," in International Semantic Web Conference, 2016.

[10] B. Shneiderman, "The Eyes Have It: A Task by Data Type Taxonomy for Information visualizations," in The Craft of Information Visualization. Elsevier, 2003, pp. 364–371.

[11] R. Johnson et al., "The Spring Framework–Reference Documentation," Interface, vol. 21, 2004, p. 27.

[12] J. J. Carroll et al., "Jena: Implementing the Semantic Web Recommendations," in Proceedings of the 13th international World Wide Web conference on Alternate track papers & posters. ACM, 2004, pp. 74–83.

[13] C. Ogbuji, "SPARQL 1.1 Graph Store HTTP Protocol," W3C REC, vol. 21, 2013.

[14] M. Bostock, V. Ogievetsky, and J. Heer, "D^3 Data-Driven Documents," IEEE transactions on visualization and computer graphics, vol. 17, no. 12, 2011, pp. 2301–2309.

E-mergency

Towards an Uberized Emergency Medical Service

José Augusto de Oliveira Neto, Nadja Maria da Silva de Brito
Center of Biologic and Applied Sciences
Dentistry Department
UEPB – State University of Paraíba
João Pessoa, Paraíba, Brazil
e-mail: zedeguga@gmail.com, nadjamso@gmail.com

Luciano Barbosa de Oliveira, Renata de Souza Coelho Soares
Post-Graduation Program on Healthcare and Technology
UEPB – State University of Paraíba
Campina Grande, Paraíba, Brazil
e-mail: barbosa.lano@gmail.com, drarenatacoelho@gmail.com

***Abstract*—This work presents E-mergency, a cloud-based software solution designed to respond most recurrent issues of Emergency Medical Services (EMS). The solution is being designed and used by means of cooperation with a public EMS responsible to answer emergency call from a one million inhabitants population and the first main functionalities have been deployed to use.**

Keywords-Emergency Medical Services (EMS); mobile application; uberized services; SAMU.

I. INTRODUCTION

Medical Emergency Services (EMS), commonly referred to as Ambulance Service, date from the 19th century [1]. In this secular history, this service became part of day-by-day routine of people and healthcare professions in multiple countries and continents. During this time, considerable improvements have also been adopted applying innovative technologies and medical procedures to pursue higher effectiveness of the service, expressed in numbers of patient life savings and cost-effectiveness of the service provisioning.

Despite the enormous differences from the first EMSs and the current high-technology vehicles and advanced trained healthcare team, some serious issues remain challenging healthcare managers and professionals to deliver a fast, efficient and cost-effective service to thousands, hundred thousand and some cases millions of requiring citizens.

In this work, we focus on four of the most recurrent issues of EMS provisioning and propose an innovative technology–centered solution to address them. E-mergency is designed to re-engineer the current EMS classical provisioning model so as to overcome critical issues and improve its service cost-effectiveness.

Section II presents a broad scenario of EMS in Brazil and the current technology resources in use. In Section III, the problem addressed in this work is stated, while in Section IV the proposed solution is detailed. In Section V the first results reached are discussed, followed by a brief conclusion and future works on them in Section VI. References can be found in the end of the paper.

II. CONTEXT

EMS in Brazil is named *SAMU* – Mobile Urgency Assistance Service - and is a public and free service, nationally funded by the Ministry of Healthcare and locally operated by thousands of municipalities all over the country.

SAMU's assistance follows well-known [2][3][4] international standards and medical protocols, in which an ambulance is required by a citizen to a call center, commonly by means of a phone call to a widely publicized phone number, where the assistance starts from the moment the operator answers the call and ends, in the worst case, when the patient is transferred to the attention of a hospital emergency team.

In the last decade, the SAMU coverage has increased and most recent official data show that since 2017 the service is available to nearly 80% of Brazilian population [5] in their home municipality (see Table 1).

TABLE I. EVOLUTION OF SAMU POPULATION COVERAGE

Population Coverage		
Year	***Population***	***Percentage***
2012	***135.703.665***	***70.53%***
2013	***141.089.175***	***72.73%***
2014	***150.487.160***	***74.84%***
2015	***155.983.958***	***76.91%***
2016	***157.299.697***	***76.92%***
2017	***163.590.587***	79.36%

According to Brazilian healthcare legal definition [6] the service is available when the following components are provided to the municipality inhabitants as depicted in Figure 1:

- A call center to answer emergency healthcare assistance phone requisitions;
- Ambulance/s vehicle/s and healthcare professionals to answer the requisition;
- Hospital/s to receive the patients assisted by SAMU.

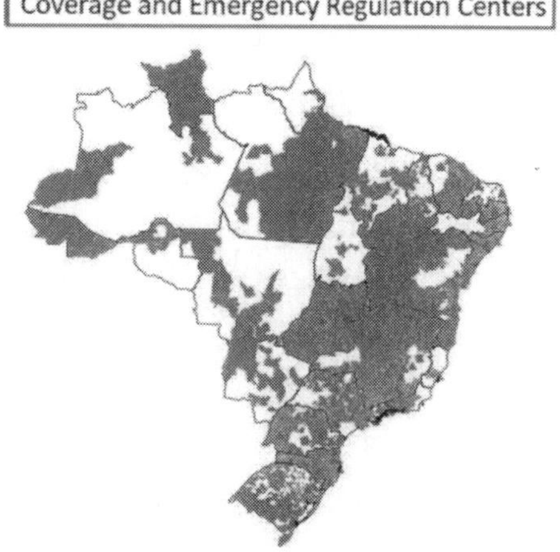

Figure 1. SAMU geographic coverage.

The typical assistance flow goes through the three components above, and starts with call center answering an assistance requisition, to which an ambulance + healthcare team is assigned to assist. When a hospital intervention is needed, a hospital bed must be reserved to receive the patient during and after the emergency room assistance.

Intense information production and exchange occurs during the assistance protocol. The mean, the time frame and the precision in which such critical information is generated, recorded and passed on can make great difference to both the patient wellbeing and the emergency service cost efficiency

Currently, phone, radio devices, desktop computers and paper forms are the dominant instruments used by citizens and SAMU professionals to produce, to record and to exchange information throughout the assistance protocol.

III. PROBLEM STATEMENT

Different studies [7][8][9] indicate that each step of the assistance protocol previously described faces difficulties to be properly executed. Some of these difficulties are expressive to Brazilian SAMU, some are equally found in emergency services from different countries and continents. Throughout the assistance protocol steps the following issues are highlights [10]:

1. High rates of fake calls;
2. Too long time interval from the emergency call to the care scene;
3. Imprecision/incompleteness of patients' healthcare records;
4. Ill ambulance-hospital communication.

Information and Communication Technologies (ICT) solutions have been placed in use to address the issues above [11][12]. Emergency support software systems have made important steps to reinforce the quality and agility of emergency services, such as EHR (Electronic Healthcare Record) data integration to access patient's medical data during the assistance protocol [13].

Although existent ICT solutions that EMS have reached some consistent outcomes, they have not been enough to effectively address the issues 1-4 (section III) and, therefore, such services still lack of technology support to make the expected quality shift.

In section IV, E-mergency is presented. An Uber-inspired ambulance ICT solution that aims to contribute to a new service model redesigned to best respond to current challenges of emergency healthcare services.

IV. E-MERGENCY

Emergency is a cloud-centered ICT solution, composed by three modules, plus a web service standard interface, which supports the different actors who require and provide EMS. wo of the component modules are mobile applications (apps), one to be used by the citizens, the other to be used by the ambulance team. The third and core module is in charge of service management and is operated in the local emergency healthcare service headquarter. Further details on each module are provided next (Figure 2).

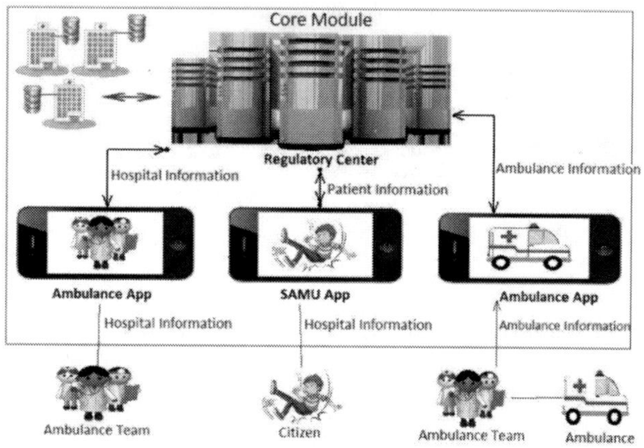

Figure 2. Emergency overview.

A. *The SAMU App*

Instead of memorizing an emergency number, citizens are now supposed to have the SAMU app installed in their mobile devices so as to make an one touch call to an ambulance, in the presence of a critical situation.

Through a simple interface (a single tap, a single button) as shown in Figure 3, the ambulance call is made and the user can then begin to interact with the emergency service.. Voice, audio messages, text and videoconference are available options to the citizen interact with the EMS call center. After this first contact, when an ambulance is assigned to the request, the same citizen can then interact with the ambulance team and forward precious information to the an agile response in the care scene. The citizen can also trace the ambulance path, which provides geographic position and expected time wait.

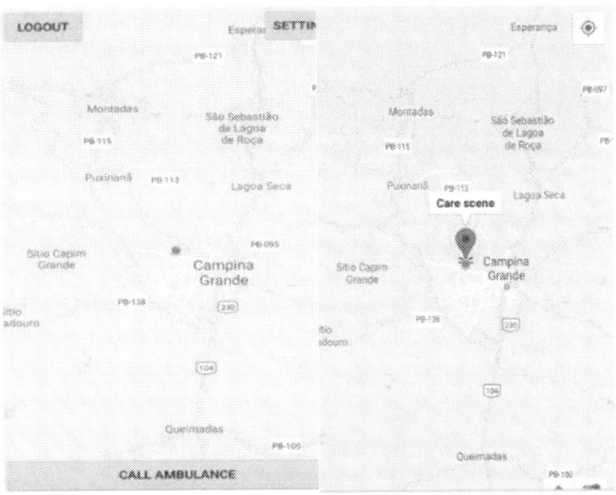

Figure 3. SAMU and Ambulance apps.

When installing SAMU app the citizen is required do provide personal information, photo, a valid official document and telephone numbers. When calling an ambulance, the app informs the EMS the citizen's GPS location. During the interaction citizen-ambulance team real time picture and video of the citizen can also be required. This set of identification and authentication functionalities addresses and is expected to strongly discourage fake calls to the emergency service (issue 1, section III).

B. The Ambulance app

The ambulance team makes use of an app to first respond to a broadcast or specific and directed request sent by the emergency service headquarter.

From the moment the ambulance answers the request, the ambulance app access all the information related to the citizen call. The type of emergency, GPS positioning of the care scene and citizen contact (in case interaction is necessary) are made available to the ambulance team (Figure 2).

During and after the assistance in the care scene, the patient healthcare information is input by the ambulance team using the app. This information will be transferred to the hospital emergency service, in the cases that the patient condition requires hospital assistance. Because the patient information is input in the app according to the assistance medical protocol and the healthcare professional mostly selects options in a clean app interface, instead of handwriting, the patient information records tend to be more detailed and precise (issue 3, section III).

The app is also the contact link between the ambulance team and hospital emergency service. From the moment that the EMS decides that hospital assistance is required.

C. The Core Module

The core and most complex E-mergency module is a cloud accessed application that manages the ambulance and hospital resources so as to best respond the citizens' ambulance request.

Answering the citizens' calls is the most use-intense function of the core module. This function puts the citizen in contact with an attendant and records all the information the citizen first provides on the emergency situation.

Is by means of the core module that the EMS headquarter transmits a request to the ambulance/s in order to assign one of these vehicles to the citizen request. The transmission can be a one-to-one call directed to a specific ambulance selected by the emergency service by means of any rationale, such as location or equipment compliance. The transmission can also go broadcast to be answered by a group of ambulances in similar conditions to answer a request.

The third and fundamental function of the core model is the interaction with hospital/s. When either the call center attendant in the first contact or later the ambulance team during local or remote assistance decides that the situation requires hospital care, the core model consults among the chain of hospitals associated to the EMS, which has the most appropriate condition to receive the patient currently assisted by the emergency service. Most appropriate here refers to the hospital medical capabilities and the estimated transfer time from the care scene to the hospital emergency room. Consults means an automated contact between the E-mergency core model and the hospitals internal ICT solution used to offer and allocate hospital bed and medical teams.

The combination of the SAMU app immediate contact with the emergency service, the Uber-similar mechanism to assign an ambulance to the citizen request and e prompt and automated interaction with the hospital is expected to efficiently respond to the (issue 2 and issue 4 , section III).

D. The Hospital Interface

E-mergency is designed to interact with hospitals in two moments. The first when a hospital bed and emergency team is allocated to receive the patient initially assisted by the emergency ambulance professionals. The second during the ambulance team assistance and patient transfer, when patient medical records are transmitted by the ambulance staff to the hospital.

Both interactions mentioned above are meant to be automated and do not entail human communication. The information on hospital availability to receive the patient and the bed and team allocation is 'negotiated' between E-mergency core module and a hospital software (ICT solution), while transmission of patient medical records to the hospital software is an interaction between the Ambulance app and the hospital software.

To make automated communication between E-mergency and hospital possible and scalable to a large number of hospitals and ambulances, one standard interface is defined by E-mergency.

Each hospital is required to implement and make available the hospital-E-mergency interface in order to become associated to the EMS The interface is a set of functions that allows to the necessary data exchange and software interoperability involved during the automated interaction between the hospitals and the emergency service.

V. First Results

The design and implementation efforts to produce E-mergency have great part of partnership articulation, to get in touch with end-users and management levels in the EMS and hospital. Currently it is established a solid cooperation involving E-mergency development team, the local emergency service, and one of the associated hospital. Such partnership has been critical to design the solution, to execute test scenarios and collect feedback on the modules deployed.

The SAMU app, the ambulance app and some of the core module have been delivered, with the following functionalities

The SAMU app

- Call function;
- Ambulance tracing;
- Text interaction with the ambulance.

The Ambulance app

- Answer to an ambulance call;
- Text interaction with the citizen.

The Core Module

- Answer to an emergency call;
- Ambulance requisition (one-to-one and broadcast)
- Hospital bed request.

Hospital interface

- Fully defined;
- Answer to bed request implemented.

VI. Conclusion and Future Work

Ambulance/Emergency is a challenging healthcare service required and provided all over the world. Some classical issues remain unsolved despite the constant improvement through ICT use in different activities of the service.

In this work an innovative approach and solution was presented aiming to apply well established technology and service provisioning model to address EMS recurrent issues.

E-mergency has its first functionalities delivered and is expected to be in full operation in the first quarter of 2019 in a one million inhabitants real use scenario. The service will be put in experimental operation coexisting with the Campina Grande SAMU regular service, with the purpose to collect comparative data on the assistance provided by the different approaches.

In the second round of development, a more innovative service model is proposed. An E-Mergency two modules only solution will be tested, in which there is no core Module in operation and the citizen SAMU request is captured directly by the ambulance located closer to the care scene. In this model the citizen SAMU App interacts directly to the Ambulance App (and team). The objective is to save costs of an EMS call center and the time to respond the request, while not jeopardizing the assistance quality standards.

References

[1] New York Presbiterian Hospital, available at https://www.nyp.org/safety/emergency-medical-services/ambulance-service-history. [retrieved: june, 2018].

[2] National Model EMS Clinical Guidelines, NAMSEMSO Medical Directory Concil, [Online] Available at: https://www.nasemso.org/Projects/ModelEMSClinicalGuidelines/documents/National-Model-EMS-Clinical-Guidelines-2017-Version2.1-29June2018.pdf. . [retrieved: june, 2018].

[3] T. Krafft, et al. European emergency data project (eed project)—ems data-based health surveillance system 3-980-8099-5-1, Brandt GmbH, Köln (2006).

[4] Ontario Provicial Ministry of Health and Long Term car, Ambulance Service Communicable Disease Standards, August 2015, [Online] available at: http://www.health.gov.on.ca/en/pro/programs/emergency_health/docs/ehs_amb_srvc_comm_disease_standards_v2_en.pdf, . [retrieved: june, 2018].

[5] Brazil, M. of Healthcare d., 2017a. DATASUS Information Technology at the Service of SUS. [Online] Available at: http://tabnet.datasus.gov.br/cgi/tabcgi.exe?cnes/cnv/leiintbr.def [retrieved: august, 2017].

[6] Brazil, M. of Healthcare (MH). Instructive Handbook to the Chain of Urgecy and Emergency Assistence Services in Braziliana Unified Healthcare Sistem (SUS). Brasília: Publisher MH.

[7] S. Lee, "Role of parallelism in ambulance dispatching," IEEE Transactions on Systems, Man, and Cybernetics: Systems, vol. 44, no. 8, pp. 1113–1122, 2014.

[8] C. S. Lim, R Mamat and T. Braunl, "Impact of ambulance dispatch policies on performance of emergency medical services," IEEE Transactions on Intelligent Transportation Systems, vol. 12, no. 2, pp. 624–632, 2011.

[9] A. Rowlands, "An evaluation of pre-hospital communication between ambulances and an accident and emergency department," Journal of Telemedicine and Telecare, vol. 9, no. 1, pp. 35–37, 2003.

[10] M.C. Reddy, S.A. Paul, J. Abraham, M. McNeese, C. DeFlitch and J. Yen, "Challenges to effective crisis management: Using information and communication technologies to coordinate emergency medical services and emergency department teams," International Journal of Medical Informatics, vol. 78, no. 4, pp. 259–269, 1 abr. 2009.

[11] Y. Shen, J. Lee, H. Jeong, J. Jeong, E. Lee, and D. H. C. Du, "SAINT+: Self-Adaptive Interactive Navigation Tool+ for Emergency Service Delivery Optimization," IEEE Transactions on Intelligent Transportation Systems, vol. 19, no. 4, pp. 1038–1053, 2018.

[12] J.L Wiler, C. Gentle, J.M Halfpenny, A. Heins , A. Mehrotra, M.G. Mikhail and D. Fite, "Optimizing Emergency Department Front-End Operations," Annals of Emergency Medicine, vol. 55, no. 2, pp. 142–160.e1, 2010.

[13] P. Wang, Z. Ding, C. Jiang and M. Zhou, "Design and implementation of a web-service-based public-oriented personalized health care platform," IEEE Transactions on Systems, Man, and Cybernetics Part A:Systems and Humans, vol. 43, no. 4, pp. 941–957, 2013.